Quantification and the Quest
for Medical Certainty

Quantification and the Quest for Medical Certainty

J. Rosser Matthews

PRINCETON UNIVERSITY PRESS

PRINCETON, NEW JERSEY

Library of Congress Cataloging-in-Publication Data

Matthews, J. Rosser, 1964–
Quantification and the quest for medical certainty /
J. Rosser Matthews.
p.   cm.
Includes bibliographical references and index.
ISBN 0-691-03794-9
1. Clinical trials—History.   2. Medicine—Research—Statistical
methods—History.   I. Title
[DNLM: 1. History of Medicine, Modern.   2. Clinical trials—
history.   3. Probability. WZ 55 M439g 1995]
R853.C55M38 1995
619′.09—dc20
DNLM/DLC
for Library of Congress   94-24091

This book has been composed in Utopia

Princeton University Press books are printed on acid-free paper and meet the
guidelines for permanence and durability of the Committee of Production
Guidelines for Book Longevity of the Council on Library Resources

Printed in the United States of America

1   3   5   7   9   10   8   6   4   2

*This book is dedicated to my mother
and to the memory of my father*

———————————————

CONTENTS

ACKNOWLEDGMENTS

As is often the case with one's first academic book, this manuscript had its origin as a doctoral dissertation which, in this case, was completed at Duke University and awarded through the history department. As such, my most immediate scholarly and intellectual debts are to members of my doctoral committee for training me. For his knowledge of the history of science as well as for his critical and editorial comments on innumerable drafts (both as a dissertation and in the subsequent revision to a book), I acknowledge my doctoral supervisor Seymour H. Mauskopf. In addition, I would like to express my appreciation to the historians of medicine Michael R. McVaugh and Peter C. English for their comments on this study at various points in its evolution. I would also like to thank Allan M. Brandt for agreeing to be on my final doctoral defense committee and raising important issues that I hope I have been able to incorporate in this revised manuscript.

Various other teachers over the years have provided intellectual stimulation that, no doubt, had an impact on the views I express in this book. Among those I would like to single out specifically in this regard are Lloyd S. Kramer and Orest Pelech for their courses in graduate school which introduced me to some of the broader issues involved in attempting to understand (and write about) the history of "modern thought."

The intellectual inspiration for this study derives from the books, articles, and dissertations that have appeared in recent years on the history of probability and statistics by such contributors as Ian Hacking, Lorraine J. Daston, Stephen Stigler, Donald MacKenzie, Andrea A. Rusnock, Ulrich Tröhler, Harry M. Marks, the contributors to the two-volume work *The Probabilistic Revolution*, and Theodore M. Porter. I would like to thank Theodore M. Porter specifically for his comments, which helped transform this manuscript into a book.

Several libraries and archives have provided indispensable help in my research on this book. Among the library staffs that have been the most helpful are the interlibrary loan office of Perkins Library at Duke; Barbara Busse and Gayle M. Elmore of the Trent Collection in the history of medicine in the Duke University Medical Center Library; the National Library of Medicine, Bethesda, Maryland; and the Library of the College of Physicians, Philadelphia,

Pennsylvania. I would like to acknowledge the following archives for permission to quote from papers housed in their collections: the American Philosophical Society—Raymond Pearl papers; the library of the London School of Hygiene and Tropical Medicine—Ronald Ross papers; and the Library, University College London—Francis Galton and Karl Pearson papers. In addition, I would like to acknowledge Major Greenwood's grandsons Roger M. Greenwood and John M. Greenwood for their permission to quote from their grandfather's correspondence in these various archives. I have made a "good faith" effort to locate the literary heirs of the other individuals whom I cite; however, so far I have been unable to locate any of them.

Several of my friends from graduate school have provided the intellectual, social, and emotional support that is indispensable in the production of scholarly research. Among the individuals who must be singled out are George Ehrhardt, Nancy Zingrone, and Carlos Alvarado.

Finally, I dedicate this book to my mother, Barbara M. Matthews, and to the memory of my father, John R. Matthews, Jr. (1920–1992). It is hard to express my degree of appreciation to them except to say that, if I have succeeded academically, the reason is that there were two people in the world who have always believed in me.

Any errors that remain are, of course, my own fault.

Quantification and the Quest
for Medical Certainty

INTRODUCTION

ONE OF THE MOST OBVIOUS contemporary examples of the effects of statistically based science both on public and on private life is the clinical trial. Since World War II, the trial has evolved into a standard procedure in the introduction of new drugs in most of the major Western industrial democracies. Its features include the use of a so-called "control" group of patients that do not receive the experimental treatment, the random allocation of patients to the experimental or control group, and the use of blind or masked assessment so that the researchers do not know which patients are in either group at the time the study is conducted. Enshrined in public policy since the 1970s,[1] the clinical trial nicely illustrates the desire of modern democratic society to justify its medical choices on the basis of the "objectivity" inherent in statistical and quantitative data.[2]

Even though the clinical trial has moved to the forefront of medical research only within the past generation, the use of comparative statistics in a therapeutic context has a much longer history, which this book will illuminate. My primary focus is three debates over the use of comparative statistics in a medical context, namely the dispute between the clinicians Risueño d'Amador and P.-C.-A. Louis before the Paris Academy of Medicine in 1837, the dispute between the mathematician Gustav Radicke and research physiologists conducted in the pages of the *Archiv für physiologische Heilkunde* during the 1850s, and finally the early-twentieth-century dispute between the British biometrician Major Greenwood and the bacteriologist Almroth Wright over the latter's technique for diagnosing disease by measuring the opsonic index. My reasons for organizing this study around these three debates are multifaceted.

One obvious benefit of comparing debates that were carried on in different countries and in different time periods is the opportunity for both cross-national and cross-temporal comparison.[3] Furthermore, the center of gravity of this study shifts to each country (France, Germany, Great Britain) at precisely the moment in time when that country was in a period of medical and/or scientific ascendancy. I focus on debates among early-nineteenth-century Parisian clinicians as they were attempting to forge (in Michel Foucault's words) the "birth of the clinic."[4] In like manner, I focus on debates among German physiological researchers when they were attempting to create the modern physiological research institute. Finally, I focus on a British context wherein Karl Pearson and his associates at University College London gave birth to what became, in effect, the first modern department of mathematical statistics. This study thus not only illuminates important debates over the use of a statistical methodology within a medical context; it also uses these debates to cast certain epochal transformations within the history of Western science and medicine into especially bold relief.

By analyzing how pioneering clinicians, physiologists, and bacteriologists viewed the "novel" method of statistical comparison, this study sheds new light on medical research and practice as a social and cultural activity[5] and demonstrates that, despite the very real differences in outlook between the medical practitioner and the medical researcher, both share an antipathy toward methods of quantitative or statistical inference. For both the medical practitioner and the medical researcher, the amorphous concept of "medical judgment" (whether that judgment is executed at the bedside or the laboratory bench) cannot be reduced to a set of explicit rules. In short, medical judgment is a form of "tacit knowledge."[6]

In addition to showing how these now-classic issues in the philosophy of science were debated in a historical context, the controversies over the use of statistical and probabilistic reasoning within medicine can also be embedded in a growing scholarly literature called the Probabilistic Revolution. This literature has dealt with such subjects as how the concept of chance became part of our conceptual vocabulary over the past two centuries and the relationship between probability and new standards of precision, argument, and objectivity.[7] By showing how these concerns have been addressed in a specifically medical context, this study extends the analysis of a "probabilistic revolution" to a wholly new domain, namely the realm of medical therapeutic research.

The central concern of the first two chapters is the aforementioned debates before the Parisian Academies of Science and Medicine in the 1830s regarding what the Parisian clinician P.-C.-A. Louis (1787–1872) chose to call the numerical method, i.e., the attempt to determine therapy on the basis of statistical comparison. Louis maintained that the use of such a numerical method would confer scientific status on the clinical physician. His principal critic, Risueño d'Amador, maintained that medicine was an "art" rather than a "science." The capstone of this debate came with the 1840 publication of *Principes généraux de statistique médicale* by Jules Gavarret, a former military engineer turned physician. Gavarret showed how to apply mathematical and probabilistic considerations to clinical statistics;

however, his approach was rejected even by Louis's supporters. They were more concerned with using numerical results to demonstrate empirical facts than to engage in probabilistic theorizing.

Even though Gavarret's approach was never systematically put into practice during the course of the nineteenth century, his ideas were discussed by various nineteenth-century commentators, as Chapter Three demonstrates. Social statisticians (this study focuses largely on Anglo-French commentators) downplayed Gavarret's mathematical concerns as introducing needless technical abstraction into otherwise clear statistical evidence. They were more concerned with demonstrating the essentially lawlike character of widespread social phenomena such as the spread of disease. Chapter Three concludes by focusing on how German clinicians and epidemiologists differed from Anglo-French contemporaries in their attitude toward Gavarret. Reflecting their strong institutional ties to the emerging research university, German commentators tried to modify Gavarret's formulas so that medical therapy could become both "scientific" and practicable.

Chapter Four studies two related issues: the legacy of Louis's "numeral method" and how physiologists defined their research agenda in opposition to Louis by claiming that experimentation on living organisms in the laboratory (rather than merely collecting statistics in the clinic as Louis had done) was the key to providing medicine with a scientific foundation. The clearest exponent of this view was the French experimental physiologist Claude Bernard (1813–1878). When the physiologists did make generalizations about their experimentally determined data, they relied on their professional medical expertise, what they called the "logic of facts." This view of the physiologists that they were uniquely qualified to interpret physiological data is illustrated concretely at the end of the chapter through an analysis of the Radicke-Vierordt debate.

The shift from statistics as a broadly descriptive tool for social analysis to a method of mathematical and scientific inference was inaugurated by the pioneering work of Francis Galton (1822–1911) in establishing what Ian Hacking has called the autonomy of statistical law.[8] This idea is central to the fifth chapter of this book. As a result of his researches into heredity, Galton pioneered the concept of "correlation": the degree to which any two series of numbers were correlated would be represented as a percentage from 0 to 1. Galton found a willing disciple of the concept of correlation in Karl Pearson (1857–1936), a professor of applied mathematics at University College London. In the context of furthering the study of heredity, Pearson created what became, in effect, the first department of modern statistics conceived as an applied mathematical discipline. Among Pearson's students was the physician Major Greenwood (1880–1949), who later furthered this mathematical and scientific vision of statistics in the milieu of medical research by holding posts successively at the Lister Institute for Preventive Medicine, the Medical Research Council, and the chair of Epidemiology and Vital Statistics at the newly created London School of Hygiene and Tropical Medicine.

In this part of the book, I have augmented the printed primary sources by studying the private correspondence of the principal players involved. The reasons for introducing archival evidence in the analysis of the third debate are numerous. In the two earlier debates over statistical methodology, the technical or mathematical issues had always been introduced by "outsiders," or figures marginal to mainstream medical thought. In the highly stratified world of post-Napoleonic French higher education, Gavarret had the unique experience of having been trained both at the Ecole Polytechnique and as a physician. Similarly, Radicke was a mathematician and meteorologist by profession, not a physician. Thus, these debates were isolated events rather than the basis for a medical or statistical research program. In Greenwood's career, however, one sees the first attempt to create the mathematically trained medical statistician as a new professional role within the world of medical research. Also, Greenwood attempted to tie mathematical statistical methods to the domain of laboratory-based research rather than the predominantly social concerns of earlier generations.

The sixth chapter continues to focus on the attempts by Greenwood and his American counterpart Raymond Pearl (1879–1940) to create the new "scientific" professional role for the statistician in the medical research establishment. Among the various strategies they deployed were writing textbooks, publishing in scientific journals, training students, and, in the case of Greenwood, writing historical narratives on the role of statistics in medicine. One of Greenwood's students who would eventually succeed him both at the Medical Research Council and the London School of Hygiene and Tropical Medicine was Austin Bradford Hill (1897–1991). In 1946 Hill designed for the Medical Research Council what is generally recognized as one of the first "modern" clinical trials to determine the effect of streptomycin on tuberculosis. In the trial Hill used the principle of randomization in assigning patients to the experimental or control groups, thereby displaying an incorporation of elements of chance into a "scientific" experiment.

Chapter Seven focuses on the efforts of Hill and his associates during the 1950s and early 1960s to convince the medical profession at large of the enormous utility of introducing the double-blind clinical trial as a standard procedure for determining the efficacy of new drugs. During this period Hill and his followers published textbooks and anthologies, attempted to offer instruction in statistical methods, and planned scholarly conferences on the role of statistical methods within medicine. These proselytizing efforts for a statistical methodology were of undeniable importance in laying the groundwork for the triumph of the clinical trial; however, the catalyst that eventually brought the clinical trial to the forefront of public and medical debate was the Thalidomide scare of the early 1960s. This scare led to the institutionalization of the clinical trial as the ultimate arbiter in determining the efficacy of experimental drugs in most of the Western industrial democracies. Thus, I argue that the clinical trial eventually triumphed less because of internal debate within the medical profession and more because society at large decided that, in an era of highly potent industrially produced drugs, the decisions of the medical profession would have to be regulated.

In the Conclusion I speculate on how this historical study can provide a framework for understanding contemporary debates about the clinical trial, both political and academic. Politically, the design and implementation of the clinical trial offers a unique opportunity for engaging some of the most hotly contested issues of our day: race, class, gender, the AIDS epidemic, and even abortion. Academically, the "objectivist" rhetoric of the supporters of the clinical trial makes it an ideal topic for those engaged in the ongoing multidisciplinary attempt to "rethink objectivity."[9]

PROBABLE KNOWLEDGE IN THE PARISIAN SCIENTIFIC AND MEDICAL COMMUNITIES DURING THE FRENCH REVOLUTION

IN THE TWILIGHT of the Enlightenment, the last years of the Old Regime in France, Louis XVI appointed A.R.J. Turgot (1727–1781) to the post of Controller-General in 1774 to help deal with the impending financial crisis. Turgot was a friend of the philosophes. He was committed not only to solving particular social and financial problems but also to realizing the possibility of a "science of man" based on reason and experience rather than tradition. His appointment proved decisive in forging strong links between the leading scientific thinkers of the day and the government that desired to use their expert knowledge—what Charles Gillispie has characterized as a union of "science and polity."[1]

One of the first problems that Turgot took on was public health. To deal with a cattle plague that emerged in November of 1774, he established a commission on epidemics headed by Felix Vicq d'Azyr (1748–1794), the personal physician to Marie Antoinette and a member of both the Académie Royale des Sciences and the Académie Française. Within two years, this commission evolved into the Royal Society of Medicine, with Vicq d'Azyr as permanent secretary. This new society helped pioneer the collection of statistical data on matters of epidemics, epizootics, and meteorology. Although it was abolished during the French Revolution in 1793, it became the model for successor institutions such as the Royal Academy of Medicine in the nineteenth century.[2]

In addition to using his position in the government to advance the cause of public health, Turgot was also interested in questions of constitutional reform. He envisaged reordering the French state on a representative basis. Village assemblies would elect district assemblies; district assemblies would elect provincial assemblies; and provincial assemblies would then elect a national assembly. One of the principal legacies of Turgot's vision was the inspiration of the Marquis de Condorcet (1743–1794), a mathematician and Secretary of the Parisian Academy of Sciences, to apply his mathematical training to the problem of the reform of elected assemblies. In his 1785 work

Essai sur l'application de l'analyse à la probabilité des décisions rendues à la pluralité des voix, Condorcet applied the mathematical theory known as the calculus of probabilities to voting patterns of judicial tribunals. He attempted to determine the probability that the judgment rendered by a tribunal was "correct."[3]

Although Condorcet was motivated to undertake his study by his particular association with Turgot, his belief that a relationship existed between the calculus of probabilities and legal or constitutional issues can be seen as a function of broader concerns within eighteenth-century probabilistic thinking. As Lorraine Daston has cogently argued, the calculus of probabilities was less a theory than a domain of applications throughout the eighteenth century. These applications were often framed in terms of the legalistic notions of equity and expectation as formulated in fair contracts. Daston sees these legal notions as central to the emergence of probability theory because there had existed in continental legal thought a system of so-called legal proofs that associated circumstantial evidence with certain numerical fractions in order to determine the likelihood that the charged individual was guilty.[4]

The principal intellectual inspiration for Condorcet's association of probability with legalistic concerns was the posthumously published 1713 work of Jakob Bernoulli, the *Ars Conjectandi*. In addition to spelling out various methodological rules for determining whether an individual was guilty or not, Bernoulli's work was significant because it developed a theory of inverse probability. This theory declared that the frequency of an event would begin to approach its probability of occurrence provided the number of observations was sufficiently large. On the basis of Bernoulli's work, Condorcet saw the possibility of extending the calculus of probabilities beyond games of chance to practically all areas of human judgment. He declared that Bernoulli "seemed to recognize more clearly than anyone the full potential of the applications of this calculus, and the manner in which it could be extended to almost all questions subject to reasoning."[5] Condorcet saw both the physical and social sciences as related in a kind of graduated skepticism. In his acceptance speech before the Academy of Sciences in 1782, Condorcet had declared: "In meditating on the nature of the moral sciences, one cannot indeed help seeing that, based like the physical sciences upon the observation of facts, they must follow the same methods, acquire an equally exact and precise language, attain the same degree of certainty."[6]

Condorcet's specific conclusions about how to apply probability theory were less important than the role he played in directing his younger and more mathematically proficient colleague Pierre-Simon

Laplace (1749–1827) to take up probability theory. Laplace had come to Paris in the 1760s as a young man and soon received the patronage of the mathematician and philosophe Jean Lerond d'Alembert (1717–1783) on the basis of his mathematical aptitude. Beginning in the 1770s, Laplace published a series of papers on various aspects of mathematical probability ranging over both theoretical mathematical issues and practical applications such as demography and vital statistics. In his popular work *Essai philosophique sur les probabilités* (1814), a collection of ten popular lectures originally given at the Ecole Normale in 1795, Laplace argued for the wide domain of possible applications for the calculus of probabilities:

> Strictly speaking it may even be said that nearly all our knowledge is problematical; and in the small number of things which we are able to know with certainty, even in the mathematical sciences themselves, the principal means for ascertaining truth—induction and analogy—are based on probabilities; so that the entire system of human knowledge is connected with the theory set forth in this essay.[7]

Laplace's belief that probability theory could be used to overcome the insufficiencies of the human intellect in potentially all domains of analysis has been attributed by Daston to his acceptance of the psychological views of Etienne Condillac. According to Condillac's empiricist epistemology, all ideas are derived from sense experience. Thus, in a well-ordered mind, judgment would have a kind of mathematical precision that the theory of probability could capture; subjective belief would be given mathematical and objective order.[8] As Laplace put it, probability theory was "only common sense reduced to a calculus."[9]

Laplace explicitly cited medical therapy as a possible domain for applying the calculus of probabilities. He advised physicians that the preferred method of treatment "will manifest itself more and more in the measure that the number [of observations] is increased."[10] Laplace cited inoculation as a procedure that was justified on the basis of numerical comparison. He even called the pioneer of the smallpox vaccine, Edward Jenner, "one of the greatest benefactors of humanity."[11]

Other mathematical popularizers of the theory of probability emphasized as well that the "calculus of probabilities" was essentially direct comparison when applied in a therapeutic context. In the 1833 edition of his *Traité élémentaire du calcul des probabilités*, Sylvestre François Lacroix (1765–1843) observed that it was by "the comparison of the number of these successes, with the total number of maladies treated following each method, and the maladies abandoned to nature ... that the relative merit is assigned to these methods."[12]

This view that the calculus of probabilities was synonymous with direct comparison of therapeutic successes and failures was accepted by both supporters and critics of Laplace within the medical community. Particularly prominent commentators who held opposing views were Pierre-Jean-Georges Cabanis (1757–1808) and Philippe Pinel (1745–1826), two physicians important in the founding of the Paris clinical school. Cabanis received his medical training at the University of Montpellier, an institution noted for holding vitalist views in theoretical medicine and cautious Hippocratic clinical observation in practical medicine.[13] In his association with the group of thinkers known as the Auteuil circle, Cabanis came into contact with Condorcet. He admired Condorcet's efforts to attain certainty and shared his general belief that knowledge was derived from experience and expressed in a clear language. Like Laplace, Cabanis maintained that the belief that a particular drug was an effective remedy for a particular pain "becomes more uncertain in proportion as the instances of similar success, in similar cases, are less numerous." For Cabanis, it was only "by repeated observations in different circumstances, that it [the therapy] comes to attain a very high degree of probability."[14]

Despite his philosophical sympathy with the views of Condorcet and Laplace on the usefulness of the "calculus of probabilities," Cabanis's medical background contributed to a skepticism about adopting such a comparative procedure as a standard practice in diagnosis and therapy. Fearing that such a quantitative approach would be a distraction, he declared in his *De degré de certitude de la médecine* (1788) that "in medicine, all, or almost all depends on ... a happy instinct, the certitude that is found most often in the sensations similar to that of an artist."[15] In a work published seventeen years later, Cabanis referred in a footnote to Condorcet and "the excellent lesson of his colleague Laplace" but still maintained that medicine "cannot be reduced to the calculus."[16]

This notion, that diagnosis and therapy should be based on a kind of informed-professional judgment rather than quantitative precision, derived from a view of medicine that emphasized the principle called specificity. According to this view, the proper professional behavior for the physician in diagnosing and treating disease was to match the idiosyncratic characteristics of individual patients with their environments. The physician acquired knowledge more through the concrete practice of medicine, which enabled him to judge individual cases in all of their uniqueness, than on the possession of quantitative or esoteric knowledge, which he could apply only according to certain fixed rules of method.[17]

Despite the prevalence of the view that each individual patient was unique, the method of direct comparison, or the "calculus of probability," did find support from Cabanis's colleague Philippe Pinel. Pinel was born in Languedoc in 1745 and received his initial training from the faculty of medicine in Toulouse. He went to Paris in 1778 and was eventually admitted to the Auteuil circle through his friendship with Cabanis. Pinel decided to focus his medical practice on the treatment of the insane and attempted to ground his treatment in scientific understanding to "bring it back to the general principles of which it is destitute."[18] One of the tools Pinel used to base his diagnosis and treatment on a "scientific" foundation was quantitative thinking.

Pinel embarked on his quantitative project around 1800, when he received at the Salpêtrière the patient population of curable insane people transferred from the Hôtel-Dieu.[19] On the basis of his researches, Pinel concluded that one could use the "calculus of probabilities" to determine the effectiveness of various therapies by counting the number of times a treatment produced a favorable response. If a treatment had a high success rate, one could generally assume that it was effective. Although Pinel alluded to the work of the mathematician Daniel Bernoulli on inoculation in the middle of the eighteenth century, there is little evidence that Pinel meant more by the phrase "calculus of probabilities" than accurate record keeping. He summarized his work by saying that "the fundamental principle of the calculus of probabilities will always be an easy and simple application when one has acquired a distinct knowledge of the respective number of events, favorable and contrary."[20]

The contrasting views of Cabanis and Pinel regarding the "calculus of probabilities" are noteworthy in that they foreshadow the types of issues that would emerge in subsequent debates over probability theory within the Parisian medical profession during the 1830s. Cabanis rejected quantitative reasoning as an intellectual distraction. He viewed medicine more as an "art" than as a "science."[21] Pinel, in contrast, made the claim that medical therapy could attain the character of a true science only "by the application of the calculus of probabilities,"[22] even though his main introduction to probability theory had been popular accounts. His work showed no detailed knowledge of the theory's mathematical subtleties as they were concurrently being developed by Laplace in his more sophisticated writings.

Parisian physicians, in general, would have little opportunity to learn about the more technical aspects of probability theory in the early years of the nineteenth century; medical and mathematical training continued to be separate despite the reforms in both scientific and medical education following the French Revolution. The principal institutional home for advanced mathematical thinking was the Ecole Polytechnique, which had been designed to train future civil and military engineers in the mathematical and physical sciences. It provided a means of livelihood for such theorists of probability as Laplace and his student Siméon-Denis Poisson (1781–1840). Also, the medical and life sciences were undergoing transformation. Medicine and surgery were united with an emphasis on anatomical knowledge, and training became centered in the various Parisian hospitals and clinics.[23] As a result of this institutional separation of mathematical and medical training, most physicians who appealed to the "calculus of probabilities" were, in fact, merely arguing in favor of arithmetical comparison of average values. The use of the phrase "calculus of probabilities" served more of a rhetorical function, by permitting the clinical physician to argue that his conclusions possessed "scientific" validity. In the Parisian medical world of the 1830s, the individual most associated with this attempt to create a science of medicine through the use of numerical comparison was the clinician Pierre-Charles-Alexandre Louis (1787–1872).

LOUIS'S "NUMERICAL METHOD" IN EARLY-NINETEENTH-CENTURY PARISIAN MEDICINE: THE RHETORIC OF QUANTIFICATION

THE QUESTION of how the clinical physician could acquire scientific credentials was of considerable importance in the medical environment in which Pierre-Charles-Alexandre Louis worked, the Parisian hospital ward. As Erwin Ackerknecht has observed, the new hospital-based training for the physician (institutionalized by the Revolutionary convention in December 1794) placed emphasis on physical examination, the use of autopsy, and statistics.[1] In recent scholarship on this new medicine, however, much greater attention has been lavished on the combination of clinical observation with pathological anatomy as revealed through autopsy (what Russell C. Maulitz has called "morbid appearances"[2]) than on the role of statistics (and hence Louis). Although the reasons for this are multifaceted, probably one of the most important was that pathological anatomy, rather than clinical statistics, captured the imagination of the French philosopher Michel Foucault. In his *The Birth of the Clinic*, Foucault declared: "Life, disease, and death now form a technical and conceptual trinity. The continuity of the age-old beliefs that placed the threat of disease in life and of the approaching presence of death in disease is broken; in its place is articulated a triangular figure the summit of which is defined by death."[3]

Another equally important reason why Louis's approach has not received as much attention as that of contemporaries who were also trying to forge a new "science" of medicine was that Louis's "numerical method" was not innovative from a methodological standpoint. It consisted of little more than the direct comparison of average values between competing therapies. Also, more recent scholarship has greatly undermined the long-dominant historiographical tradition that saw Louis as the pioneer in the use of numerical comparison to determine the efficacy of medical therapies;[4] examples have been found of eighteenth-century medical practitioners who similarly argued for therapeutic innovations based on quantitative comparison.[5]

Even though Louis's originality has recently been questioned, his great historical significance is in no way undermined. Rather, we gain the opportunity to view his contributions in the broader medical and scientific context of his time, in which Louis emerges as one of the myriad of often-competing voices in the Parisian medical world of the early nineteenth century who argued that medicine must be grounded in a scientific foundation.[6] However, unlike the others, Louis maintained that the physician could become scientific without leaving the clinic. His contemporary physiologically minded colleagues (François Magendie being the most famous) had begun to look to the laboratory as a way of creating a science of medicine autonomous from both clinical practice and government regulation.[7] Thus, Louis did not appeal to the prestige of science to make medical research distinct from medical practice. Rather, his appeal to a discourse of "science" reflected the continuing legacy of the Enlightenment project in a postrevolutionary context. Even those engaged in clinical practice should have their judgment informed by the tools of science.[8]

Several factors in the social milieu of the Parisian medical and scientific establishment contributed to Louis's view that enumeration was synonymous with scientific reasoning. Not least among these was the widespread popularity of Laplace's *Essai philosophique sur les probabilités* of 1814, which had gone through five editions before its author's death in 1827. The popularity of this work has been attributed by Ivo Schneider to its aims, which accorded well with the meritocratic principles emphasized in the new systems of education and training instituted in the first half of the nineteenth century.[9] Laplace addressed these issues of the potential for personal advancement through education in the final sentences of his essay:

> If we consider the analytical methods to which this theory has given birth; the truth of the principles which serve as a basis; the fine and delicate logic which their employment in the solution of problems requires; the establishments of public utility which rest upon it; the extension which it has received and which it can still receive by its application to the most important questions of natural philosophy and the moral science; if we consider again that, even in the things which cannot be submitted to calculus, it gives the surest hints which can guide us in our judgments, and that it teaches us to avoid the illusions which ofttimes confuse us, then we shall see that there is no science more worthy of our meditations, and that no more useful one could be incorporated in the system of public instruction.[10]

Louis, in particular, found in Laplace's admonitions a justification of his method, which consisted of careful observation, systematic record keeping, rigorous analysis of multiple cases, cautious generalizations, verification through autopsies, and therapy based on the curative power of nature. In his studies number was used for diagnostic purposes.[11] The focus, however, remained on concrete empirical results rather than abstract or mathematical theorizing.

This empirical vision of science can be seen as a function of Louis's training as a clinical physician. He had studied at Rheims after switching careers from law to medicine, completing his training in Paris in 1813. After a three-year stint in Odessa, Russia, Louis returned to Paris to study childhood diseases, particularly diphtheria. Eventually, Louis accepted an appointment at the hospital *La Charité*, where he collected records for six years on therapy and pathology.

Based on cases that he observed at *La Charité* in the last three months of 1821, Louis argued in favor of the numerical method in his *Recherches anatomico-pathologiques sur le phthisie* (1825). He declared that "the edifice of medicine reposes entirely upon facts, and that truth cannot be elicited, but from those which have been well and completely observed." The introduction of number into diagnosis and therapy would ensure that all medical practitioners arrive at identical results: "[a]ll is not then obscure or uncertain in medicine, when the observations which guide us are exact."[12]

Louis applied the same kind of empirical method in his study of typhoid fever. From observations collected between 1822 and 1827, he determined that it produced morbid alteration in the large intestines. He concluded, "Thus it is that facts confirm facts, and when one conclusion has been rigorously deduced, every circumstance which can be referred to it, is a new proof of the truth of it."[13]

Louis observed that typhoid fever was a disease of the young since the mean age of the 50 cases who died was 23 and, of the 88 who recovered, 21. In commenting on this result, Louis spelled out his view of the relationship between the numerical method and uncertain knowledge:

> The slight difference, observed between the mean age of the patients, who died and of those who recovered, is not owing to *chance* [my emphasis], for the reader will perceive that before the age of twenty-five years, the number of those who recovered is much greater than that of those who died, whilst after that age the number of both is almost the same. So that if youth is a necessary condition for the development of the typhoid affection, this affection is so much less formidable according as those who experience it are younger.[14]

In addition Louis attempted to document the claim that the length of residence in the city of Paris reduced an individual's chance of contracting or dying from typhoid fever. The results of Louis's findings are in the following two tables[15]:

TABLE 1

Mortality and Residency in Paris

10	had been at Paris from	2 to 3 weeks
8	" " " " "	3 to 5 months
10	" " " " "	6 to 10 "
9	" " " " "	11 to 20 "
5	" " " " "	20 to 30 "
2	" " " " "	4 to 8 years

44 total

TABLE 2

Recovery and Residency in Paris

7	had been at Paris from	2 weeks	to	3 months
19	" " " " "	3 months to	5 months
19	" " " " "	6 " " 10 "
20	" " " " "	11 " " 20 "
12	" " " " "	20 " " 30 "
1	" " " " "	30 " " 40 "
7	" " " " "	4 to 8 years

85 total

Finally, Louis addressed the issue of bloodletting as a therapeutic method to combat typhoid fever; he observed that 39 out of the 52 fatal cases had been bled. The mean duration of their disease was 25.5 days as opposed to 28 days for those who were not bled. Louis maintained that "at first view, venesection would seem to have accelerated the fatal progress of the disease."[16] Of the 88 patients who recovered, 62 were bled with the mean duration of the disease being 32 days as opposed to only a 31-day mean duration for those not bled—"a first result which little favors the action of bleeding."[17] Clearly, Louis was not yet ready to dismiss bloodletting. The average number who were not bled that recovered was not appreciably different from the average number who were bled that recovered—the two series differed by only 1 in 14. Louis concluded:

Thus, whatever view we take of the facts, we see, in blood-letting, a therapeutic agent of some utility, in the course of the typhoid affection, when we employ it properly, and at a period near the commencement of the disease; and this agreement in the results ought to give them a degree of importance to which the small number of facts on which they rest would not seem to entitle them.[18]

Louis did not remain convinced that bloodletting was "a therapeutic agent of some utility." In his 1835 treatise *Recherches sur les effets de la saignée*, he produced what was to become one of the most famous applications of the numerical method as well as one of the most sustained critiques of bloodletting as a medical therapy. He observed that bloodletting had an effect on the progress of pneumonitis; however, the effect was much less than previously supposed. Also, he noted that pneumonitis was never arrested in its early stages by bloodletting and he determined that age exerted a great influence on the rapidity of progress of diseases. Finally, he declared that antimony administered in large doses could substitute for bloodletting and that vesication had no evident influence on the progress of pneumonitis.[19]

Throughout the work the discussion was motivated by an attempt to disprove therapeutic assumptions through an appeal to empirical "facts." For instance, Louis argued against bloodletting because 18 patients died out of the 47 who were bled—approximately 3 out of every 7—while there were only 9 deaths out of the 36 patients not bled, producing a lower mortality rate of approximately 1 in 4.[20] To show the effect of bleeding on angina tonsillaris, 23 cases were used and 13 were bled. The average length of the disease for those not bled was 9 days as compared to 10-1/4 days for those who were bled. For Louis, the differences "can only be attributed to the employment, or omission of bloodletting."[21]

The fact that Louis had chosen to focus on the therapeutic procedure of bloodletting was in itself significant. This method of treatment had been vehemently defended by François Joseph Victor Broussais (1772–1838), the chief physician at the Parisian military hospital and medical school, the Val-de-Grace. In his 1816 work *Examen de la doctrine médicale généralement adoptée*, Broussais argued in favor of bloodletting on the basis of his new doctrine of "physiological medicine." Diseases were to be identified by observing the lesions of organs. Treatment would consist of local bleeding of the diseased organ and low diet, since most diseases were the result of inflammation, especially of the gastrointestinal tract. Bloodletting was part of a more general program to guarantee that the practices of the clinician possessed "scientific" validity.[22]

Broussais's work provides a broader medical context for the clinical studies of Louis. By rejecting bloodletting and the theory of physiological medicine on which it was based, Louis was, in effect, offering an alternative vision of what constituted "scientific" reasoning within clinical medicine. The clinician would acquire scientific credentials by employing aggregative thinking about a population of sick individuals rather than using pathological anatomy to observe disease in a particular individual. Louis observed that human beings acquire knowledge because "we can form a class of facts bearing sufficient resemblance, one to another, and from hence deduce laws which every day's experience verifies."[23]

By placing emphasis on quantitative thinking at the level of the social group rather than on qualitative understanding at the level of the individual patient, Louis was attempting to appeal to the authority of number to justify clinical judgment. He contended that the difference between numerical results and words such as "more or less," and "rarely or frequently" is "the difference of truth and error; of a thing clear and truly scientific on the one hand, and of something vague and worthless on the other."[24] With the triumph of the numerical method:

> we shall hear no more of medical tact, of a kind of divining power of physicians. No treatise whatsoever will continue to be the sole development of an idea, or a romance; but an analysis of a more or less extensive series of exact, detailed facts; to the end that answers may be furnished to all possible questions: and then, and not till then, can therapeutics become a science.[25]

By drawing on popular expositions of the calculus of probability in texts such as those of Laplace, Louis's could invoke the authority of mathematical reasoning. He noted that enumeration reduced rather than augmented the inevitable errors in human judgment:

> [I]t is impossible to appreciate each case with mathematical exactness, and it is precisely on this account that enumeration becomes necessary; by so doing the errors, (which are inevitable,) being the same in the two groups of patients subjected to different treatment, mutually compensate each other, and they may be disregarded without sensibly affecting the exactness of the results.[26]

He maintained that he had made himself "master of the spirit of mathematical science" by using the calculus of probability.[27]

Louis's appeal to the calculus of probability was mainly a rhetorical move to demonstrate that the clinical physician could employ quantitative methods; he showed no interest in studying the theoretical models actually developed by his mathematical colleagues.

His only statement on what in present-day parlance would be the issue of sample size or statistical significance was that, in an epidemic, a treatment was effective after it had been tested on 500 patients because "among so large a collection, similarities of condition will necessarily be met with, and all things being equal except the treatment, the conclusion will be rigorous. In this manner has the treatment of Asiatic Cholera been estimated."[28] One of Louis's students, Bruno Danvin (1808–1867), even displayed an outright hostility to the applications of probability given by the mathematicians. In a thesis submitted to the Paris school of medicine in 1831, Danvin declared:

> Compare the supporters of numerical analysis to Condorcet, in respect to his bizarre pretensions. . . . I ask what type of relationship ought to be established between the abstractions of the intellectual and moral world, and the facts of which the medicine of observation is composed, facts which are before our senses, are appreciated in their manner of being, their termination, and always with respect to their organization, from which they are never abstracted?[29]

Even though Louis had invoked the mathematical theory of probability, his numerical method reflected a concern more with the collection of empirical facts than with abstract theorizing.

Louis's work had, however, framed the issue of quantification in clinical medicine in the terms in which it would be debated before the Parisian Academies of Science and Medicine in the late 1830s: should the proper concern of the clinician be the individual patient or an aggregate of sick people? The triggering mechanism for these debates came from another medical context, the question of the proper surgical procedure for removing bladder stones. As Ulrich Tröhler has observed, the arguments justifying surgical procedures to remove bladder stones had been framed in terms of numerical comparisons of successes and failures from the early eighteenth century. Itinerant practitioners of the procedure of cutting to remove the stone in the urinary bladder (known as lithotomy) claimed success by appealing to figures. Since they received more medical practice if they had a higher success rate, they arranged their results in tabular form and reported both the number of patients treated and the number of deaths on a yearly basis. Extensive debate took place before the Parisian Academies of Science and Surgery throughout the eighteenth century over technical modifications of the procedure. Thus, the appeal to statistical comparison had become well established by 1835 when the surgeon and urologist Jean Civiale (1792–1867) published a report for the Academy of Sciences on a new bloodless method for removing bladder stones (lithotrity).[30]

Even though Civiale's statistical approach was not original, he collected empirical data on a much wider scale than had earlier compilers of surgical statistics. In his research he benefited from the political changes following the Revolution of 1830. His work was sponsored by the Ministry of Public Instruction, which had been put under the direction of the Sorbonne history professor François Guizot; Guizot had helped to reconstitute the Académie des Sciences Morales et Politiques in the hope that it would become a scientific advisory body of the government in all matters pertaining to the moral and political sciences.[31]

Civiale argued in a manner similar to Louis's: given the fallacy of human memory, surgeons would tend to remember their successful cases more than their unsuccessful ones; errors would emerge in results that were not recorded exactly.[32] He compared the relative rate of death from the traditional surgical procedure and the lithotrity. With the older procedure, there were 4,478 recoveries out of 5,715 operations performed (i.e., a 78.4% rate of recovery). In contrast, there were only 6 deaths out of 257 cases treated by lithotrity (i.e., a 97.7% rate of recovery).[33] For Civiale, at least, the evidence was conclusive that the lithotrity was the operation of choice.

In response to Civiale's statistical results, the Academy of Sciences established a commission including the mathematician Siméon-Denis Poisson and the clinical physician François Double (1776–1842); the latter was the principal author of a report that appeared on October 5, 1835. Double's educational and philosophical outlook was similar to that of Cabanis. He had studied at Montpellier, earning his degree in 1799. Subsequently, he had been appointed to the editorial board of what became the *Journal général de la médecine* and had been a founding member of the Academy of Medicine.[34] In his report to the Academy of Sciences, Double spoke as one of the leaders of the Parisian medical establishment.

Double used his report on surgical statistics to engage the broader issue of the proper function of the medical profession within society as a whole. He was heir to an older tradition that emphasized the primacy of a humanitarian ethos. The principal function and cultural authority of the physician derived from his unique ability to diagnose disease and alleviate individual human suffering. In criticizing the numerical method, Double was rejecting the attempt to turn the clinician into a scientist through the use of aggregative thinking and quantification.

In contrast to Louis's emphasis on a group of sick individuals, Double believed that the proper concern of the physician should remain the individual patient. To apply the numerical method, it would be necessary "to strip away the individual in order to arrive at the elimination of all which the individual would have been able to introduce accidental to the question."[35] Double maintained that medical diagnosis was more of an art; it was not appropriate to "elevate the human spirit to that mathematical certainty which is found only in astronomy."[36] Although framed in terms of a discussion of quantification, the central issue turned on the question of contrasting visions of professional legitimacy for the clinical physician (i. e., humanitarian healer or empirical scientist).

The fact that a debate over the proper structure of clinical discourse (quantification) could be seen as a function of social and professional self-definition was not fortuitous. As Thomas Laqueur has recently argued, the literary structure of the clinical case history from the eighteenth through the early-nineteenth century could be seen as an instance of the "humanitarian narrative." The case history constituted step-by-step accounts of the suffering of a particular human being; it was designed to make real the pain of others and to offer a logic of specific intervention.[37] By advocating quantitative thinking for clinicians, Louis seemed to be implicitly questioning this sense of a humanitarian professional ethos to which physicians such as Double adhered.

Despite this difference in attitude toward quantitative thinking, both Double and Louis framed their discussions of whether clinical medicine should aspire to "scientific" status by invoking the phrase "calculus of probabilities," which had first appeared in Laplace's popular *Essai philosophique sur les probabilités* (1814), and they used it as an umbrella term to designate any conclusion expressed in numerical terms. For other members of the Parisian scientific establishment (such as those with mathematical training from the Ecole Polytechnique), however, the phrase referred to a branch of mathematical reasoning useful for analyzing errors of observation in astronomy or geodesy; it connoted a vision of science that emphasized mathematical precision more than the empirical judgments of clinical physicians. This difference in meaning was evident when Double's report was criticized by Claude Navier (1785–1836), a former student at the Ecole Polytechnique and an engineer at the Ecole des Ponts et Chaussées. He conceded to Double that results could be determined by experience and induction rather than by the calculus of probabilities. Nevertheless, he maintained that "the employment of the calculus of probabilities gives to this method [induction] the necessary rigor and exactitude." Navier concluded:

> It is not thought . . . that in applying statistical procedures and the calculus of probabilities to medicine, that the results thus obtained can be considered as theorems of geometry. It is thought only that more precision can be given to consequences deduced from observations, and that the application will be rendered less uncertain.[38]

Although remaining cordial, Double responded from the perspective of the practicing physician, affirming that "the eminently proper method in the progress of this science [medical diagnosis] is logical analysis and not numerical analysis."[39] The disagreement nicely highlighted how the mathematically trained and the clinical physicians could share a vocabulary even though they did not understand each other.

Since the issue of these debates over the "calculus of probabilities" was whether or not clinical medicine should remain an "art" or aspire to "scientific" credentials, it is not surprising that the subject was taken up in 1837 by the Royal Academy of Medicine, which had been created by a royal ordinance of December 20, 1820. Louis XVIII had been persuaded of the need for such an organization by his personal physician, Antoine Portal. Portal had *ancien régime* sympathies and tried to revive institutions such as the Royal Society of Medicine in forms more appropriate for the times. Like its predecessor the Royal Society of Medicine, the Royal Academy of Medicine was designed to serve both as a technical consultant to the government and as the supreme arbiter of medical matters in France.[40] As the prospectus of the first volume of its *Bulletin* asserted, the academy has "recognized the mission to conserve and propagate vaccines, to respond to the demands of the government, regarding epidemics, endemics, epizootics, on different cases of legal medicine and public hygiene, on the value of new remedies, or secret remedies, [and] on the properties of natural minerals."[41] The question of whether medicine should aspire to become scientific was clearly germane.

Once again, the structure of the ensuing debate (although not its technical content) was framed around issues advanced by the mathematically trained, namely the notion of *l'homme moyen* or the "average man" of Adolphe Quetelet (1796-1874), and the "law of large numbers" of Siméon-Denis Poisson (1781-1840). In his 1835 work, *Sur l'homme et le développement de ses facultés*, Quetelet defined *l'homme moyen* as the average of all human attributes in a given country. It would serve as a "type" of the nation analogous to the idea of a center of gravity in physics. Quetelet formulated his ideas by combining his training in astronomy and mathematics with a passion for social statistics. He had been sent to Paris by the Belgian government to learn astronomical methods in 1823. During his stay, he acquainted himself with the principal mathematical texts of Jean-Baptiste Joseph Fourier, Laplace, Lacroix, and Poisson. In addition to these theoretical concerns, Quetelet collected data on aggregate phenomena like birth, death, marriage, crime, and suicide and arranged his results according to age, sex, profession, and place of residence. Quetelet then utilized his mathematical training regarding these statistical results to create what he called social physics. He developed an elaborate system of metaphors and similes linking the social domain to the theories and mathematics of physics and astronomy. Central to his view of the social order was the theoretical construct of the average man.[42]

Quetelet noted the potential usefulness of the concept of an "average man" in advancing medical diagnosis. In the preface to his work, he observed:

> All the sciences tend necessarily to the acquirement of greater precision in their appreciations. The study of diseases, and of the deformities to which they give place has shown the benefit derivable from corporeal measurements; but in order to recognize whatever is an anomaly, it is essentially necessary to have established the type constituting the normal or healthy condition. In order to be of use to science, I have deemed it necessary to direct my researches in a particular manner to the dimensions of the chest which seem most frequently to merit consideration in the state of illness; and the same region is the one where the greatest malformations are most often to be observed.[43]

Quetelet saw the presence of means as a general phenomena occurring throughout society: "In a given state of society, resting under the influence of certain causes, regular effects are produced, which oscillate, as it were, around a fixed mean point, without undergoing any sensible alterations."[44] Quetelet claimed that every quality, taken within limits, was essentially good; it is only in extreme deviations from the mean that it became bad.[45]

Quetelet observed that in healthy individuals the number of inspirations and beats of the heart are generally confined within certain limits. He cited the results of physiological researchers such as François Magendie who had reported that the number of beats of the heart at the time of birth varied between 130 and 140. Also, he noted that in general the number of inspirations per minute was 20 and that every fifth inspiration was deeper than the others. Quetelet suggested similar limits for the number of pulsations in both children and adults at different ages.[46]

Quetelet maintained that the concept of an average man was as useful to medical practice as it had been to medical research:

> The consideration of the average man is so important in medical science, that it is almost impossible to judge of the state of an individual without comparing it to that of another imagined person, regarded as being in a normal condition, and who is intrinsically no other than the individual we are considering. A physician is called to a sick person, and, having examined him, finds his pulse too quick, and his respiration immoderately frequent, &c. It is very evident, that to form such a decision, we must be aware that the characters observed not only differ from those of an average man, or one in a normal state, but that they even exceed the limits of safety. Every physician, in forming such calculations, refers to the existing documents on the science, or to his own experience; which is only a similar estimate to that which we wish to make on a greater scale and with more accuracy.[47]

For Quetelet the concepts of health and disease (whether applied to societies or individuals) could be represented in terms of statistically constructed norms.

Just as Quetelet had envisaged a potential use for his concept of an average man in the realm of diagnosis, so likewise did the mathematician Siméon-Denis Poisson envisage a use for what he dubbed the "law of large numbers" in the realm of medical therapeutics. Poisson had had a long-standing interest in the calculus of probability. He had published an excerpt from Fourier's 1807 work and two of Laplace's works in the *Bulletin de la Société Philomatique* in 1810–1811 and also edited Laplace's monumental *Théorie analytique des probabilités* (1812). His own first work on the subject, read before the Academy of Sciences on March 13, 1820, dealt with a game of chance. Furthermore, by the 1830s, the calculus of probabilities and mathematical physics had both made advances. With the death of Jean Nicolas Pierre Hachette in 1834, who had been in charge of the course on descriptive geometry and mathematical instruments in the Paris Faculty of Sciences, Poisson proposed the creation of a chair for either the calculus of probabilities or mathematical physics. The choice to create this chair in the calculus of probabilities was made by the minister of public instruction, François Guizot, who saw that the position went to his personal friend Guillaume Libri. Poisson exchanged his course with Libri in the 1836–1837 academic year in order to write *Recherches sur la probabilité des jugements en matière criminelle et en matière civile* (1837). In this work, Poisson revived the discussion initiated by Condorcet and Laplace of applying probability theory to the voting patterns of judicial tribunals.[48]

Unlike previous theorists of probability, Poisson could utilize the annual compilation of legal statistics begun by the French Ministry of Justice in 1825 to try to determine the probability of a given individual being convicted. Poisson noted that if an event were observed a large number of times, one could assume that the probability of its occurrence in the future would increasingly correspond to the percentage of times that it had been observed in the past, provided the number of observations were sufficiently large. This "law of large numbers" was described as "the basis of all applications of the calculus of probabilities."[49] In a footnote to the preface of his work, Poisson suggested how the calculus of probabilities could be applied to medical therapy: if "the number of cases where it [a medication] has not succeeded is very small compared to the total number of cases . . . it is very probable that the medication will succeed in a new test."[50]

Poisson made the mathematical formulation of his law of large numbers explicit by observing that if an event occurred m times after μ observations, the average $[m/\mu]$ could vary within possible limits of oscillation given by $m/\mu \pm u/\mu\sqrt{2mn/\mu}$ with $n = \mu - m$ by a standard of reliability given by the relation $R = 1 - (2/\sqrt{\pi})\int_u^\infty e^{-t^2} dt$.[51] Thus, the value selected for u would determine the probability that the actual average would never vary beyond the possible limits of oscillation. As Poisson observed:

> It will always be possible to take u large enough for this probability R to differ as little as one wants from certitude. It will not even be necessary to give to u a large value to render very small the difference $1 - R$: it will suffice, for example, to take u equal to four or five in order that the exponential e^{-u^2}, the integral $\int_u^\infty e^{-t^2} dt$, and, as a result, the value of $1 - R$, will be almost imperceptible.[52]

Poisson eventually decided to select the value $u = 2$, because this produced a high standard of certainty, namely $P = 0.9953$ or 212:1.

Both Quetelet's and Poisson's work had raised the issue that aggregative thinking might provide scientific insight for clinical medicine. In this respect, their ideas were like those of Louis. Furthermore, the general dichotomy between the moral imperative to heal the individual and the desire to advance knowledge by making comparisons within a population, continued to be central to the Academy of Medicine debate of 1837, even though the issues were now explicitly framed in terms of Quetelet's "average man" and Poisson's "law of large numbers."

The debate commenced at the behest of Jean Cruveilhier (1791–1874), an opponent of the use of statistical methods and the first individual in France to hold a chair devoted to the study of pathological anatomy.[53] In response to Cruveilhier, the Montpellier professor of pathology and general therapy Benigno Juan Isidoro Risueño d'Amador (1802–1849) requested an audience before the academy. At the session of April 25, 1837, he delivered a lecture on the role of the calculus of probabilities and its application to medicine.[54]

Risueño d'Amador began by declaring that the question under discussion was "the grand question of certitude in medicine."[55] Following in the Montpellier tradition of Cabanis and Double, he contended that the principal concern of the physician should be healing the individual sick person. The theory of probability was "only a curious logical abstraction."[56] It was true a priori; however, real decisions must be made on the basis of experience, i.e., a posteriori.

Risueño d'Amador used the example of maritime insurance to illustrate why he regarded the calculus as not applicable to medicine. If 100 vessels perish for every 1,000 that embark, one would still not know which particular ships would be destroyed. This depended on unknown variables such as the age of the vessel, the experience of the captain, or the condition of the weather and the seas. Analogously, the calculus of the mathematicians "cannot be used to forecast a determined event, but only to establish the probability of a certain numerical proportion between two classes of possible events. But it is precisely this fact that makes it completely useless in medicine."[57] The physician would be more concerned with determining which individual would become sick than with establishing statistical patterns of disease within a population.

Risueño d'Amador's focus on the individual patient was not only a matter of intellectual predilection but also of professional ethics. Since the calculus of probabilities could never reveal which particular individual would suffer, several would have to die of a malady before the statistical results had "proved" the efficacy of a particular therapy.[58] Risueño d'Amador also cited the arguments of the political economists Adam Smith and J.-B. Say, who had declared that statistical results were too uncertain because they merely told of results at a particular instant in time for particular individuals. Generalization of the results to future cases was impossible because of the uniqueness of the individuals involved.[59]

Risueño d'Amador was not against drawing analogies from past experience to guide in medical practice; what he denied was that these inferences could ever be represented in the mathematical formalisms of the calculus of probabilities. The physician was like an artist. Although an artist knows the general features of human appearance, he still must paint a particular individual rather than a statistical composite. Similarly, the physician knows the general appearance of disease conditions; however, he must still treat the individual patient. Echoing (and rebutting) Laplace, Risueño d'Amador declared: "The calculus of probabilities has been called good sense reduced to a calculus, but . . . one can perhaps ask whether good sense is calculable the same as intelligence, the passions, the human affections, . . . and all that which pertains to moral and intellectual life and affects human beings."[60] For Risueño d'Amador, the results of observation in medicine are often more vague than in other sciences because the things observed have a more indecisive character, not because of any failure in the observing physician. He must observe a disease in various manifestations—a problem with which the astronomer or physicist does not have to contend. Any errors in medical observation derive "from the observation itself, not from the observers; the defect is in the object, not in the instruments of our researches."[61]

Risueño d'Amador's discussion produced a heated debate among several members of the Academy of Medicine. My book highlights the views of three representative individuals: Frederick Dubois d'Amiens (1797–1873), François Double, and P.-C.-A. Louis. Dubois d'Amiens had become a physician in 1828 and a member of the Academy of Medicine in 1836.[62] He maintained that Louis's numerical method produced results no more accurate than the empirical observations of other physicians—even though Louis had dutifully displayed his results in tabular form. As Dubois d'Amiens sarcastically observed, "The formulas that he [Louis] has deduced from his calculus have all been as vague, all as approximate as those employed before him."[63] Representing results statistically did not necessarily beget greater precision.

Dubois d'Amiens admitted that one could determine a series of mean values by performing a large number of dissections and observing the anatomical variations in order to create a kind of "anatomical average man." However, such an exercise was futile because it would result in "a fictive being, a being of reason, and far from having an exact idea of that which has to be found in dissections, one will have a pattern which is never found to model exactly in nature precisely because it will always be an assemblage of *anatomical means*."[64] Even in dissection the focus of the clinical physician remained on the individual corpse rather than on the quantifiable group.

When Double addressed the Academy of Medicine, he reiterated the arguments that he had already presented in his Academy of Science report two years previously. He reaffirmed that the proper concern of the profession of medicine should remain the treatment of the individual patient. On the potential usefulness of a Queteletian average man, he observed that such a construct would reduce the physician to "a shoemaker who after having measured the feet of a thousand persisted in fitting everyone on the basis of the imaginary model."[65] On the issue of Poisson's increasingly controversial attempts to mathematize human decision-making,[66] Double responded that the pressing and immediate concerns of medical practice made such "intellectual" or "academic" debate futile:

> Since the men of mathematical certainty dispute among themselves, the physicians also tolerate not being in agreement.... The matters that doctors embrace have far more serious importance than the matters with which the geometers occupy themselves. It is necessary to agree, it is not without reason that the public is scandalized and irritated by our dissension.[67]

Double decried both the construct of an average man and a "law of large numbers" for shifting attention from the individual needing treatment to the population of sick people as a whole.

Double was not concerned with the issue of the mathematical cogency of the arguments advanced by mathematicians from Condorcet to Poisson. Rather, he maintained that medical practice did not require such quantitative analysis. Double declared that analogy and induction were "no less useful and no less certain" than the calculus. Turning Laplace on his head, Double downplayed the importance of the calculus of probability because it was "founded only on good sense."[68]

Even P.-C.-A. Louis attempted to distance his work from the views of the mathematically trained such as Quetelet. In his discussion before the academy, Louis maintained that numerical analysis was not designed to determine such an imaginary being as the average man but rather was intended as an aid to knowledge and as a tool to eliminate error in matters of diagnosis. He appealed once again to the authority of quantitative reasoning, declaring that numbers were more precise than vague adverbial expressions such as *often* or *rarely*.[69]

Such debates on the role of number in medicine did not take place solely before the Academy of Medicine but were represented within other medical literature of the time. The issue remained whether medicine should follow the quantitative methods of the physical sciences or the more intuitive methods of the so-called moral sciences. For the professor of clinical medicine in the Parisian medical faculty, Jean Bouillaud (1796–1881), ironically a student of Broussais and a supporter of bloodletting, the numerical method was commendable because it might enable diagnosis to attain "the same exactitude that is in the other branches of natural philosophy."[70] Bouillaud's associate T.-C.-E. Auber (1804–1873), however, thought that medicine should not become quantitative because it was essentially different from the natural sciences:

> Compared to other sciences, it resembles some of them, and contrarily has many differences with certain other ones; thus, for example, it has a much greater analogy with the moral, metaphysical, religious, and political sciences, while it differs essentially from the physical, chemical, and mathematical sciences, by the nature of its laws as well as by the nature of its spirit.[71]

Throughout these debates, the phrase "calculus of probabilities" had been a recurring motif. This derived more from the conviction of Louis and his followers that clinical medicine needed "scientific" status than from a desire to utilize the technical features of the calculus then being developed by the mathematicians. This disjunction was nicely highlighted by the response of the medical profession to Jules Gavarret's probabilistic approach as presented in his 1840 work *Principes généraux de statistique médicale.*

Gavarret had a unique educational background that enabled him to discuss both the medical and mathematical issues involved. He had entered the Ecole Polytechnique in 1829 intending to become an artillery officer but resigned four years later with the rank of lieutenant in order to become a physician. Subsequently he studied with Gabriel Andral, a clinician and associate of Louis, with whom he published several papers.[72] Gavarret was thus able to analyze the debates over the "numerical method" from the perspective of his mathematical training. He claimed that what motivated him to write the study was the fact that the Academy of Medicine had "neglected completely the most important part of the question [i.e., the mathematical theory of probability]" in its debates.[73]

The first part of Gavarret's study addressed the criticisms of Risueño d'Amador. In response to Risueño d'Amador's claim that medicine should be based on induction rather than the mathematics of probability, Gavarret maintained that the mathematical relations of probability theory merely expressed the statistical results of inductive reasoning in a more formal and exact manner. In response to the objection that the calculus of probabilities did not inspire much confidence, Gavarret made the standard philosophical distinction that the only truths one knows with absolute certainty are those of mathematics; all other forms of empirical knowledge are based on a certain degree of probability. Hence, it was not that the foundation of medicine was less secure than those of other sciences, but rather that all sciences rested on the same probabilistic foundation. Finally, in response to the objection that the individuality of each patient precluded the application of results contained in statistically established relationships, Gavarret gave the standard retort that all empirical sciences use generalization and if the use of analogy on the basis of similar past experience were precluded from human judgment, "medical experience would be a word devoid of sense."[74]

Gavarret conceded that if a physician wished to study a disease condition in an individual patient, the proper method of analysis should be pathological anatomy and there would be no need for numerical reasoning. However, if one desired to classify into different categories the diseases experienced by groups of individuals, one had to resort to record keeping, i.e., numerical results. In this nosological context, Gavarret declared that advances in pathological anatomy were, in fact, a necessary precondition for the application of numerical methods because they had permitted diagnosis to attain a much higher degree of perfection. In an attempt to show how numerical methods had evolved out of recent medical advances, Gavarret asserted that "it is because the important work of pathological anatomy had not yet shown to them how to separate the maladies neatly, that the ancients were generally unable to make good statistics."[75] Following in the tradition of Laplace and Poisson, Gavarret maintained that the calculus of probabilities was the only language "well enough made" both to permit ideas to be replaced by abstract numerical symbols and to derive conclusions.[76]

Although Gavarret supported the numerical method, he did not spare criticism of the relative mathematical ignorance of its other supporters. He emphasized that statistical results were accurate only if certain conditions prevailed—namely, the facts must be similar or comparable, there must be greater confidence in large as opposed to small numbers of observations—and the results derived from statistics were true only within certain limits of oscillation. Gavarret cited Poisson's formula that if an event occurred m times after a total of μ observations, then the average m/μ would vary between limits given by $m/\mu \pm u\sqrt{2mn/\mu^3}$ with $m + n = \mu$. The value selected for u determined the probability that the average would not vary outside these limits according to the formula $P = 1 - (2/\sqrt{\pi})\int_u^\infty e^{-t^2} dt$. Gavarret emphasized that the choice for u was "extremely delicate." However, he maintained that "we can rely in this regard upon an authority in whom no one ... will attempt to dispute the importance." Then Gavarret cited Poisson's assumption that $u = 2$, which resulted in

$P = 0.9953$ or 212:1. Gavarret declared that such an assumption was "necessary in order that the formulas can serve to interpret the results of all statistics."[77] Noting the general lack of awareness of these considerations among other supporters of the numerical method, Gavarret observed that it was by believing in the "absolute truth of their numerical reports ... that the partisans of the numerical method have arrived at the most contradictory results."[78]

Gavarret singled out the conclusions of Louis, in particular, as based on insufficient data. Louis had observed 140 cases of typhoid fever with 52 deaths and 88 recoveries. From these results he had concluded that the effectiveness of treatment was 37%, i.e., 52 / 140. However, on the basis of Gavarret's formula for the limits of oscillation possible after 140 cases, the results could vary by 11.55%. In other words, the death rate could vary between 26% and 49% in every group of 140 cases observed. Gavarret queried sarcastically, "What would [the results] have been if we had submitted to a similar examination the conclusions that were deduced from the twenty-three observations to establish the treatment of throat tonsillitis?"[79] In general, Gavarret advocated several hundred observations before a result could be established by the numerical method.

Gavarret also proceeded to show how the law of large numbers could actually be used in certain cases to verify claims made by Louis. Louis had declared that the preferred therapy could be determined in an epidemic by administering one therapy to 500 patients and another therapy to another 500 patients and observing which produced the lower mortality rate. On the basis of Louis's suggestion, Gavarret proposed a hypothetical experiment. For the first medication, one assumed that $m = 100 =$ the number of deaths, $n = 400 =$ the number that recovered, and $\mu =$ the total number observed. For the second medication, in contrast, the following results were assumed: $m' = 150$, $n' = 350$, and $\mu' = 500$. The difference in the relative percentage mortality for the two therapies would then be given by $m'/\mu' - m/\mu = 0.3 - 0.2 = 0.1$. In this instance, the formula for determining the limit of oscillation was given by $2\sqrt{2mn/\mu^3 + 2m'n'/\mu'^3}$ $= 2\sqrt{2(100)(400)/(500)^3 + 2(150)(350)/(500)^3} = 0.07694$. Thus, since the percentage reduction in mortality proved to be greater than the limit of oscillation for the number of cases observed, the medical researcher could assume that there was better than a 99.53% chance that the reduction in mortality could be attributed to the superior therapy.[80]

Gavarret did not limit his application of Poisson's "law of large numbers" to questions of diagnosis and therapy. He saw these techniques as useful for studying large-scale phenomena such as epidemiology: "The large epidemic of cholera that ravaged France in 1832, has justified perfectly the conclusions to which the use of the principles of the law of large numbers have led us."[81] In discussing this epidemic, he focused on an 1834 report commissioned by the prefect of the Seine and the police which recorded the number admitted to the Paris hospital each day of the week for a period of 27 weeks. The commission had concluded that the intemperance of the working classes on the weekend was a principal cause of cholera because the number admitted seemed to increase at the beginning of the week, on Sunday and Monday. In the 27-week period, a total of 7,890 had been admitted on Sundays, Mondays, Wednesdays, and Thursdays after 108 days of observation. In contrast, only 5,887 were admitted on Tuesdays, Fridays, and Saturdays after a total of 81 days of observation. Gavarret criticized the commission's conclusions by appealing to the law of large numbers.

In order to apply the law of large numbers, Gavarret made several assumptions. Since the number of days of observation was different for the two series, Gavarret set up the proportion 108 : 81 :: 7890 : X, which produced the value $X = 5,918$. Thus 5,918 represented the number that would have been admitted in the first series if there had been only 81 days of observation with "the frequency of the entries remaining the same." Then Gavarret once again modified the law of large numbers so that it could be applied to this particular example— in this case, a mutually exclusive binary opposition. He noted that if the actual chance of one event occurring is p (being admitted to the hospital on Sunday, Monday, Wednesday, or Thursday) and the actual chance of the contrary event is q (being admitted to the hospital on Tuesday, Friday, or Saturday rather than some other day in the week), the equations to determine the limits of oscillation would become $p + 2\sqrt{2pq/\mu}$; $q - 2\sqrt{2pq/\mu}$. Furthermore, Gavarret observed that if the probability of each occurrence was roughly equal (i.e., you would be just as likely to be admitted on one set of days as the other if there were no "disturbing cause"), it would follow that $p = q = 0.50$ and the equation for the limits of oscillation would reduce to $0.50 \pm \sqrt{2/\mu}$. In the context of this particular example, Gavarret observed that $m = 5,918 =$ the number of patients admitted on the first set of days; $n = 5,887 =$ the number of patients admitted on the second set of days; $\mu = m + n = 11,805 =$ the total number of observations. This meant that $m/\mu = 5,918/11,805 = 0.50131$ and $0.50 + \sqrt{2/\mu} = 0.50 + \sqrt{2/11,805} = 0.50 + 0.01302 = 0.51302$ and $0.50 - \sqrt{2/\mu} = 0.50 - 0.01302 = 0.48698$. Since the average m/μ was within the possible limits of oscillation, Gavarret concluded that "no one is authorized to think that some cause had been able to intervene to render the results relative to the first group of days more frequent than those of the second group of days."[82] Even though Gavarret conceded that there might be a slight increase in the number admitted on Monday because the central bureau in charge of admission was open only for two hours on Sunday as opposed to its normal five, he still maintained that the variation "in other respects does not notably surpass the higher limit of possible error."[83]

Gavarret's work generated an initial flurry of excitement, particularly among those members of the medical profession who desired medicine to have "scientific" credentials. As an anonymous reviewer in the *British and Foreign Medical Review* observed with regard to the well-established sciences, "Their certainty and perfection bear an exact proportion to the accuracy of the observations or experiments which they employ, and the extent to which they avail themselves of calculation."[84] Medicine, however, did not live up to this lofty ideal because of the complexity of the facts involved. Also, most physicians did not have the opportunity to collect enough cases to be absolutely certain of the effectiveness of a given therapy; therefore, Gavarret's results should be used to derive the degree of reliance to be placed on any given number of observations. The reviewer concluded with the declaration that the goal to be sought was

> to make the more perfect sciences an example to our own. . . . The practical man who delights in single cases, and glories in those who deal with them, will laugh to scorn the plodding collector of facts by the hundred, and confound the principles which he draws from them with the hypotheses, of some dreaming enthusiast. He who devotes himself to the science of medicine must expect little sympathy from the mere votary of the art.[85]

Gavarret's response to the Academy of Medicine debate was also favorably summarized by Casimir Broussais (1803–1847), a professor at a military hospital and connected to the Faculté de médecine de Paris. Casimir Broussais was the son of F.-J.-V. Broussais, against whom Louis's attack on bloodletting had been addressed. In his 1840 work entitled *De la statistique appliquée à la pathologie et à la thérapeutique*, Broussais proved to be a devout disciple of his father's physiological doctrines. He delighted in showing how Gavarret's mathematical analysis could be used against Louis's results:

> The power of numbers leads Louis to overthrow some current errors and to contest some assertions of distinguished physicians. . . . The predilection in this author for the numerical method is so pronounced that

he does not bear in mind induction and analogy when they are found to be in disagreement with his numbers, and he proclaims with authority some results that are equally repulsive both to experience and to reason.[86]

Broussais began his study by recounting the debate at the Academy of Medicine. He noted objections to the numerical method, e.g., it undermined the individuality of the patient. Also, he listed the various responses of the supporters of the method. Then he discussed medical statistics in terms of the main issues addressed by Gavarret—the nature of the facts (whether they are comparable) and the number of the facts considered (the problem of the law of large numbers). He listed the formulas already noted that Gavarret had derived from Poisson and showed how they could be used to criticize Louis's results. Broussais concluded his work with the following observations:

Statistics is henceforth an indispensable instrument in the experimental method or in observation in medicine, but its application is surrounded with a crowd of difficulties. Before counting the facts, one has to begin by clearly determining their nature, in order to make a series of those which are comparable; statistics are collected in the greatest possible number, then the ratio of these numbers is established which gives the frequency of the phenomena which is subject to the operation of chance. The average represented by this frequency or this chance is at no point fixed; it is more or less variable, all the more variable when the number of facts is very small, all the less when the number is very large. But it is not enough to apply statistics to a large series of facts; it is necessary to divide this series into fractions, and to make on each of them the same operations as on the ensemble. By means of the formula of Poisson, one appreciates the degree of variability of the average. . . . Statistics in medicine is thus necessarily inevitable; its application is urged, but it is necessary to accumulate facts, but facts well observed, that is to say facts diagnosed well, very precise.[87]

Despite the existence of such favorable commentators, the French followers of Louis were not interested in Gavarret's mathematical formulations. Their rhetorical use of the phrase "calculus of probabilities" was tied more to their professional vision of the clinical physician as a scientist; they were using the methods of reasoning advocated by Laplace, the "Newton of France." They had neither the desire nor the educational background to apply the calculus of probabilities as used by mathematicians. Rather, their method was one of empiricism. It was designed to use number and "facts" to disprove the errors derived from centuries of medical theorizing.

The hostility of Louis's followers to Gavarret's concerns was brought out most strikingly by the response of the pediatrician François-Louis-Isidore Valleix (1807–1855), who was described on his death as one of Louis's "most intelligent" disciples—a man who had "served well the doctrines of his celebrated master."[88] Valleix agreed that Gavarret had shown how to apply the calculus of probabilities to medicine, but declared that such an application would "change the face of the science."[89] He emphasized that Louis's method was empirical, i.e., derived from facts. It was not subject to the mathematical abstractions introduced by Gavarret's analysis. In response to Gavarret's suggestion that several hundred observations were needed to establish the effectiveness of a given therapy, Valleix opined that "a century of continual observation often would not be sufficient."[90] Valleix was not attacking the numerical method *per se*, as had Double and Risueño d'Amador; rather, he was attacking Gavarret's formulation of the numerical method purely in terms of probability mathematics. Just as important as the number of cases observed were such factors as the intensity of the malady or its length of duration. If these factors were taken into account, a conclusion could be reached after fewer observations than Gavarret's probabilistic mathematics required.[91]

In response to Valleix's criticisms, Gavarret reiterated that those who did research in therapeutics completely misunderstood the true principles of medical statistics. He claimed that Valleix had asked him to do more than his work intended. He desired only to expound the "general principles" of medical statistics, not show its applicability to all possible branches of medicine. Gavarret challenged Valleix to study all the abstract writings on probability; if Valleix showed them to be at fault, "then and only then" would he admit his error.[92]

Needless to say, Gavarret's approach proved to be anathema to members of the medical profession such as Double and Risueño d'Amador who emphasized the "art" of treating the individual patient. One anonymous reviewer in the *Revue médicale française et etrangère* (a journal with Risueño d'Amador on its editorial staff) praised the critique of Valleix and savaged Gavarret. The author criticized the unduly "academic" nature of Gavarret's discussion. Invoking a therapeutic metaphor that had increasingly ironic overtones, the author declared that Gavarret had "bled" and "purged" both the medical experience of centuries and his own personal experience as a physician. The work was condemned as "medical heresy" that, if implemented, would "be sufficient by itself to ruin the [numerical] method." The review concluded with the observation that Gavarret's methods had the potential to "reduce the immense majority of practitioners to the condition of isolated intellectuals, destined to accept servilely all medical ideas, which would be imposed on them by the professors in our clinics, only suitable after the collection of several hundred observations."[93]

Gavarret did not pursue the issues that he had raised in his later medical writings; his work was a particular response to the debate at the Academy of Medicine rather than a program for mathematical or medical research. After publishing a series of papers with the famous clinician Gabriel Andral on the composition of the blood, Gavarret became a doctor in 1843, writing a thesis on emphysema. In the following year he was elevated to the chair of medical physics in the Faculté de médecine de Paris, offering a course entitled "General Laws of Dynamic Electricity." He continued to teach in this capacity until his retirement in 1887, producing a series of studies on the effects of temperature on the body, astigmatism, and the physical principles involved in the process of hearing. Professionally, Gavarret became a member of the Academy of Medicine in 1858, and its president in 1882. He was made inspector general for public instruction in medicine in 1879 and produced several reports in that capacity on the need for reform in medical education. In all of these endeavors, probabilistic mathematics was only a minor or nonexistent concern.[94]

Ultimately, the debate over Louis's "numerical method" was not important because Louis was the first to apply number to diagnosis and therapy (he was not), nor because probability mathematics was combined with clinical medical statistics. Rather, it was important because of the rhetorical contexts within which all of the important participants chose to frame their discourse. Supporters of the numerical method such as Louis or Bouillaud claimed that by enumerating their results, the clinical physician could become just as much of a "scientific" thinker as the herald of probabilistic methods, P.-S. Laplace. Likewise, opponents of the numerical method such as Double and Risueño d'Amador emphasized how the "calculus of probabilities" distracted the physician from his principal concern—the individual in need of medical care. Even though both sides framed their discussion in terms of the "calculus of probabilities," the debate centered on whether the main concern of the clinician should be the individual or the aggregate of sick people. The issue turned largely on a question of professional self-definition: those who saw the physician as a scientist favored enumeration while those who saw the physician as having a moral injunction to heal the sick emphasized the individual. Thus, the debates were significant as one of the first instances in which the issue of quantification had been used to frame a discussion of the value of "science" for the clinical physician.

Probability theory, however, was more than a mere rhetorical tool for an intraclinical dispute; it was also a mathematical theory developed in the seventeenth century and revived in the middle decades of the nineteenth century by Siméon-Denis Poisson. Jules Gavarret's work was heir to this mathematical tradition. It had little impact on the concerns of the clinical physicians. The training of the clinician in the social milieu of the hospital ward was radically different from the mathematical and engineering instruction at the Ecole Polytechnique. The clinicians and mathematicians could share a vocabulary without understanding each other intellectually. However, the commentaries on Gavarret's work during the second half of the nineteenth century open a unique window into differing views of statistical reasoning within American, British, French, and German medical cultures.

Chapter Three

NINETEENTH-CENTURY CRITICS OF GAVARRET'S PROBABILISTIC APPROACH

EVEN THOUGH Gavarret's probabilistic views were never systematically put into practice in the nineteenth century, they were discussed by a variety of public health, clinical, and epidemiological commentators in American, British, French, and German publications.[1] Although all cited Gavarret as the pioneer in attempting to apply probability mathematics to medical statistics, few advocated adopting Gavarret's proposals uncritically as a "standard" procedure. In general, American, British, and French commentators did little more than reproduce some of Gavarret's algebraic equations and examples; they did not discuss the theoretical underpinnings of his reasoning because they were still primarily focused on social issues of public health. German commentators, by contrast, showed much more mathematical sophistication than did other writers on Gavarret. The reasons for this were diverse: some wanted to modify Gavarret's formulas to justify some specific therapeutic innovation, while others were concerned more generally with using Gavarret's probabilistic approach to confer scientific validity on therapeutic decisions.

One of the first favorable commentaries on Gavarret appeared in the 1844 treatise by Elisha Bartlett entitled *An Essay on the Philosophy of Medical Science.* Ironically, Bartlett dedicated the work to Louis, with whom he had studied in Paris before returning to America to become a professor of medicine at the University of Maryland. In the dedication, Bartlett declared that his essay "endeavors to illustrate, to develop, and to vindicate those principles of philosophy which lie at the foundation of your own various and invaluable researchers."[2] Even though Bartlett espoused Louis's empirical vision of medical science,[3] he did not hesitate to give additional credit to Gavarret (unlike Louis's French disciples). Bartlett praised the "admirable treatise of M. Gavarret on *Medical Statistics.*"[4] On assigning credit to those who had done the most to introduce number into medicine, Bartlett observed in a footnote that "the full measure of its [the numerical method's] value was practically exhibited by Louis, and its true principles philosophically developed and demonstrated by Gavarret."[5]

Despite his praise for Gavarret's achievement, Bartlett did not give an account of the mathematical reasoning underlying Gavarret's arguments. Rather, he cited (without proof) the example that Gavarret had given of how Louis's average of 37% mortality for typhoid fever could vary between 26% and 49% since only 140 cases had been observed. Bartlett justified this omission:

> It is not necessary to the purposes of this essay, that I should enter into a full exposition and development of the principles of statistics in their application to the different branches of medical science. It is only by the aid of these principles, legitimately applied, subject to the conditions already pointed out, that most of the laws of our science are susceptible of being rigorously determined.[6]

Another commentator on Gavarret who also began his medical career by studying under Louis was the British statistician William Augustus Guy (1810–1885). However, unlike Bartlett, Guy was principally involved in public health rather than medical therapeutics. As a result, he was much more skeptical of Gavarret's mathematical approach. After studying with Louis, Guy received his medical degree from Cambridge in 1837 and was appointed professor of forensic medicine at King's College London in the following year. He was a founder of the Health of Towns Association and active in the cause of public health, writing incessantly on such varied subjects as sanitary reform, prison reform, occupational disease, insanity, and forensic medicine. Professionally, he was an honorary secretary of the Statistical Society of London from 1843 to 1868, the editor of its journal from 1852 to 1856, its vice-president from 1869 to 1872, and its president from 1873 to 1875.[7]

In light of Guy's strong institutional affiliation with the Statistical Society of London, his distrust of Gavarret's mathematical approach to statistical reasoning is understandable. The Statistical Society of London had been created by urban middle-class elites who were attempting to deal with problems concerning the environment, public health, education, and the poor.[8] Also, all of the founding members belonged to what Susan Faye Cannon has described as the "Cambridge network"; that is, they held to an inductive view of the scientific method, they opposed deductive approaches to social analysis such as those promulgated by David Ricardo, and they had faith in the unity of the scientific method and its widespread applicability to several distinct areas of knowledge.[9] As the early editors of the *Journal of the Statistical Society* observed, the main purpose of statistics was to create a "science of man"; it was "not the mere 'method' of stating the observations and experiments of the physical and other sciences." To avoid accusations of partisanship, the prospectus of the Statistical Society declared in May of 1834:

> The Statistical Society of London has been established for purposes of procuring, arranging, and publishing "Facts calculated to illustrate the Condition and Prospects of Society."
>
> The Statistical Society will consider it to be the first and most essential rule of its conduct to exclude all Opinions from its transactions and publications—to confine its attention rigorously to facts—and as far as may be found possible, to facts which can be stated numerically and arranged in tables.[10]

Guy's view of statistics remained close to the views of the society's founders. As he observed in 1865, "The science of statistics is a comprehensive science, of which 'social science' and political economy are only branches or departments."[11]

Since Guy viewed statistics more as a tool of social analysis than a method for mathematical inference, he minimized the importance of the numerical method in general and Gavarret's highly mathematical formulation of it in particular. In an early article on the numerical method published in the *Journal of the Statistical Society of London* in 1839, Guy attempted to steer a middle course between the contrasting views presented in the Academy of Medicine debate of 1837. Like Louis and his supporters, he emphasized that "the certainty of a science is exactly proportional to the extent to which it admits of the application of numbers."[12] He invoked the image of astronomy as the paradigmatic exact science that all other fields, including medicine, should strive to emulate.[13] Like Double and Risueño d'Amador, however, Guy remained hesitant to rely solely on numerical results. The numerical method was merely one of many that the physician had at his disposal. It was a kind of pedagogical tool that would "supply the want of tact by furnishing the inexperienced with accurate calculations of . . . the probable consequences of different modes of treatment."[14] Ultimately, it was the responsibility of the practicing physician to ascertain the extent to which individual cases departed from numerically determined averages.

This view that mathematical inference could not be reduced to mathematical equations became evident explicitly when Guy addressed the potential usefulness of Gavarret. In an encyclopedia article on medical statistics, he cited the algebraic formulas used by Gavarret to establish the "limits of oscillation" for medical statistics, namely, $m/\mu \pm 2\sqrt{2mn/\mu^3}$ with $m + n = \mu$. Also, he showed how Gavarret had used these formulas to demonstrate problems with Louis's derivation of conclusions from only 140 cases of typhoid fever. However, Guy was unwilling to accept the excessive number of observations that Gavarret's formulas would entail:

> in many instances we are prevented by causes too numerous to specify from bringing together facts by the hundred or the thousand, and yet, were we to reject the smaller numbers as inadmissible, we should be thrown back upon the still more loose and less trustworthy general statements from which it is the province of statistics to rescue us.[15]

Even though Gavarret had shown that averages derived from small numbers of facts are subject to error, Guy maintained that "there is always such a probability of their coinciding with ... the true averages, as to justify their employment as standards of comparison."[16]

Guy continued this type of criticism of Gavarret when he gave the Croonian lecture before the Royal College of Physicians in 1860. He conceded that Gavarret's algebraic formulas could be used if the results obtained from taking simple averages were in doubt. However, he maintained that "such applications of the pure mathematics must be very rare; and they are certainly not free from objections." Results obtained from averaging a small number of cases could generally be assumed to be accurate. It was theoretically possible, but not very probable, that such a small number of observations could be incorrect.[17] For leading members of the nineteenth-century British statistical community, a mathematical approach to statistics like Gavarret's was more of a curious intellectual oddity than a basis for medical research.

A similar lack of interest in probabilistic mathematics was evident among those members of the nineteenth-century French medical establishment concerned with aggregative thinking. Like their British counterparts, they wished to ameliorate the disastrous effects of early industrialization rather than to engage in abstract or mathematical theorizing. Louis René Villermé, one of the pioneers of sociomedical analysis through the use of number, showed no attempt to base his conclusions on the calculation of probable error—even though he saw Quetelet's text *Sur l'homme* through the press.[18] Other French physicians who advanced a statistical approach to public health proved equally indifferent to theoretical constructs advanced by mathematicians.[19] The demographer Louis-Adolphe Bertillon (1821–1883) wrote with regard to Quetelet's central notion of the average man: "The average man of each human type is an artificial entity belonging to the category of mean indices. ... It is important to remark that this average man is less ... a scientific entity than a creation of the imagination."[20] Like their British counterparts, French statisticians were more concerned with social analysis than with mathematical inference.

Nor did Gavarret's probabilistic approach find a receptive audience in the mid-nineteenth-century French mathematical community. Irenée Jules Bienaymé (1796–1878), the sole Parisian mathematician concerned with applying probability to medicine, was clearly a marginal figure within the scientific establishment. Bienaymé had studied at the Ecole Polytechnique during the two-year period 1815–1816. In 1818 he became lecturer in mathematics at the military academy at St. Cyr. Two years later he began his civil service career by entering the Administration of Finances. He became Inspector General in 1834 and an officer of the Legion of Honor in 1844. He took early retirement from his government position because of the Revolution of 1848. In that year he was temporarily appointed professor at the Sorbonne in the Calculus of Probabilities. When Gabriel Lamé (one of the individuals to whom Gavarret dedicated his 1840 text) assumed this chair in 1851, he said of Bienaymé that he, "almost alone in France, represents the theory of probabilities, which he has cultivated with a kind of passion."[21] In 1852 Bienaymé was made a member of the Academy of Sciences. His interest in statistical questions was evident during his career as an academician; he acted as referee for the Montyon prize for statistics for twenty-three years.[22]

Bienaymé expressed his view of the role of the probability calculus within medical statistics in an article published in 1840. Although Gavarret was not mentioned, Bienaymé did make passing reference to the works of Bernoulli, Bayes, Laplace, Fourier, and Poisson in his attempt to demonstrate the "proper" methods of applying probabilistic considerations to medical statistics. He pointed out the various difficulties that one had to overcome in order to apply these formulas of the mathematicians:

> The construction of new formulas proves that medical statistics is not very advanced. . . . All science desires good statistics at first. Therefore it is important that medical statistics collect large series of facts before one is sufficiently justified in applying the formulas appropriate to the particular nature of that which characterizes them. The calculus of probabilities can be applied to all things; but it is able only to accompany the investigation of facts.[23]

Bienaymé echoed the Laplacian view of probability as "good sense reduced to a calculus" and applicable to potentially all domains of human knowledge. As Bienaymé wrote in 1853, "It is well said that the very nature of the calculus of probabilities is that it treats errors of all kinds, digressions of all types, incompatibilities of observations resulting from the feebleness of the human organs, from the discordance of the statistics given which produce physically very extensive variations."[24]

Such an older view of the wide applicability of probability theory was explicitly rejected by Augustin Louis Cauchy, the recognized leader of the French mathematical community in the mid-nineteenth century, who in his 1821 *Cours d'analyse* exhorted, "Let us therefore eagerly pursue the study of the mathematical sciences without letting them extend beyond their domain; and let us not imagine that we can approach history through mathematical formulas or sanction morality with algebraic theorems or integral calculus."[25] Because Cauchy was concerned with restricting the domains of application of the probability calculus, Bienaymé's vision of applying probability to medicine would not be apt to win his favor. To most leading members of the French mathematical establishment, Bienaymé was an outsider who had been appointed to the Academy of Sciences because of his sympathies for Napoleon III rather than his mathematical or scientific competence.

Bienaymé's career serves as a nice counterpoint to Gavarret's. In 1840 both had produced isolated works that attempted to relate medical statistics to the mathematical theory of probability. Gavarret retained professional and institutional ties to the medical establishment whereas Bienaymé retained ties to the mathematical and governmental branches of Parisian professional life. However, there was no audience in either the Parisian mathematical or medical communities for such an intellectually hybrid discourse: French statisticians (like their British counterparts) were more concerned with social description than mathematical inference, and probability mathematics had been professionally marginalized to the study of games of chance and error analysis in astronomy.

Even though Gavarret's probabilistic concerns were of little importance to nineteenth-century medical statisticians, his monograph continued to be cited in the French medical literature of the second half of the nineteenth century. One early French critic of Gavarret was Jacques Raige-Delorme (1795–1887), who had studied medicine in Paris, graduating in 1819, and is remembered principally as the editor of the medical journal *Archives générales de médecine*.[26] In an 1844 analysis, he observed that Gavarret placed greater concern on the mathematical aspects of acquired medical knowledge than on its other facets, such as an ability to diagnose disease. Gavarret had committed the error of seeing all medical facts as "a unity of the same nature as that which results from removing a black or a white ball from an urn which contained several of each color." In a probably unintentional inversion of Laplace, Raige-Delorme declared that "good sense alone is able to fix the limits."[27]

Often commentators did little more than make *pro forma* references to Gavarret's work as the definitive account of how to apply probability theory to medical statistics. In his 1874 article, the medical editor Amédée Dechambre (1812–1886) began his discussion by alluding to the "famous statistics of Civiale which had been, in 1835, the subject of an important discussion at the Academy of Sciences and the point of departure for analogous debates which resounded much later at the Academy of Medicine."[28] These debates were still the touchstone for French commentators who wished to discuss the merits of statistical methods in medicine.

Similarly, when Dechambre observed that medical statistics may vary between certain limits of oscillation that depend on the number of cases observed, he declared, "I am not able to expose the mathematical principle in detail, but it is good that the reader know of it. It is still Gavarret which will serve as our guide."[29] Even after thirty-four years, the medical authors who invoked the name of Gavarret did not explain (and probably did not understand) the mathematical underpinnings of his reasoning.

Dechambre treated Gavarret's writing in largely the same way nine years later in his seven-page addition to the article on "Statistics" in the monumental 100-volume *Dictionnaire encyclopédique des sciences médicales* (1864–86). The article itself was over eighty pages in length and focused on the more traditional concerns of nineteenth-century medical statistics—vital statistics and questions of public health. Dechambre cited the standard arguments for the numerical method such as its greater precision and mentioned that the work of Gavarret was "excellent" for showing the mathematical underpinnings of the numerical method. He did not discuss the mathematical details of Gavarret's work in depth and was cautious about applying the numerical method, observing that it provided only general results and never told which treatments to apply in a particular case.[30] Statistics was more intimately linked to issues of public health than to the determination of medical therapy.

In contrast to this dismissive treatment at the hands of British and French statistical commentators, Gavarret's work was the subject of more detailed commentary in the German medical literature. This difference derived to a significant extent from the social and professional environment in which all the German commentators worked, the nineteenth-century German research-based university.

The commitment to original research had spread through the German university system with the founding of the University of Berlin in 1810 and the wave of university reforms that followed. In the medical faculty, the institutional embodiment of this research ethos was the physiological institute in which experimental laboratory investigation was carried on, often with increasingly elaborate technical apparatus to ensure repeatability and precision. Furthermore (in direct contrast to the "scientific" empiricism of a Louis in the clinic), the ideal of physiological experimentation was seen as the way for the physician to attain the proper training necessary for medical practice. Even for medical students who would never engage in research, it was assumed that exposure to research methods would produce a respect for exactitude and sound reasoning that would be carried over into medical practice. Laboratory instruction would also ensure that the practitioner would be technically competent to keep abreast of new research developments.[31] It was in the context of this attempt to reform medical training that Gavarret's probabilistic approach was discussed by various German commentators who wanted the professional standing of medicine to derive from "scientific" aptitude.

One of the first publications to comment on Gavarret's probabilistic considerations was the *Archiv für physiologische Heilkunde*, a journal whose explicit intent was to provide a scientific foundation for medicine through the study of physiology. As the editors declared in the introduction to the first volume in 1842:

> We believe that now is the time to establish a positive science from the existing material of experience, [a science] that does not seek to ground itself in the authorities but rather in the empirical evidence which allows phenomena to be understood, and which also avoids the illusions of praxis and will lead to a deliberate, certain therapy.[32]

On the question of the value of statistical reasoning within science, the editors maintained:

> Statistical results are themselves not yet science; they are nothing but material that must be examined and that must allow its own examination. The task of scientific research is to abstract general laws from the regular and consistent repetition of phenomena. This abstraction is the living process which science produces.[33]

This view of the relationship between statistical methods and scientific reasoning is understandable in light of the educational background of one of the founders and principal editors of the *Archiv für physiologische Heilkunde*—Carl August Wunderlich (1815–1877).

After his initial training at the University of Tübingen, where he had been introduced to the French medical literature of Laennec, Andral, Louis, Magendie, and Chomel, he had studied in Paris for a good portion of 1837 before eventually returning to take an academic post at the University of Tübingen in 1840. Later he openly acknowledged his debt to the writings of F.-J.-V. Broussais.[34] He had been introduced to the French medical tradition at exactly the point in time when the issue of statistical methodology was being contested the most intensely.

Wunderlich was not as strident as both the supporters and the opponents of the numerical method had been in the debates at the Parisian Academies of Science and Medicine. He admitted that, as a practical matter, statistics applied to groups might provide some insights into therapeutic advances even if they were not sufficient to render medicine "scientific": "in medical opinion, as in all judgments of the human situation ... generally only a reckoning of probability is admissible, a reckoning of probability, which is, to be sure, not orderless, but rather is able to be employed in accordance with principles and in the possession of all of the available factors."[35]

Wunderlich's journal came to serve as a prominent publication outlet for German commentaries on probabilistic and statistical methods. One of the first articles to appear was the 1854 piece by the Karlsruhe physician G. Schweig. Like the Anglo-French commentators, Schweig emphasized the general ignorance of probability mathematics within the medical profession. However, Schweig's attitude toward a probabilistic approach like that of Gavarret was clearly more favorable than the lip service provided by earlier commentators. He observed that medicine would progress considerably and important things would be accomplished if doctors better understood the foundations of the statistical method:

> But if the largest part of the medical-statistical works cannot be used, and if this state of affairs can be traced back to a lack of knowledge about statistical methods, then it is certainly time to present prevailing norms to the medical public in such a way that the particulars can be easily understood and the whole can be grasped with precision.[36]

Problems in medical statistics were treated either trivially or from a mathematical perspective that was incomprehensible; those who knew the relevant mathematics would not be aware of the needs of the physician or medical researcher.

After giving examples of the role of statistical reasoning in medicine (such as the relative percentage of male and female children that

died out of all those born in Baden in 1852), Schweig provided a theoretical discussion of mathematical probability. He cited the pioneering work of Jakob Bernoulli, the *Ars conjectandi* (1713), as well as the
1837 work of Poisson, the contents of which were characterized as "of
very great importance for a theoretical discussion of proportions."[37]
He gave the various formulations of the "law of large numbers" as
given in Gavarret's work, e.g., $m/\mu \pm 2\sqrt{2mn/\mu^3}$ with $m + n = \mu$. Unlike
the American, British, and French commentators who had merely
reproduced the algebraic formula verbatim, Schweig observed that
this equation for the limits of oscillation depended on assuming a
standard of 212:1 (0.9953) that the results were not the product of
chance. This realization that the particular form of Gavarret's algebraic expressions was a function of the standard that had been assumed would become the foundation of all subsequent German
commentaries.

Schweig concluded by observing that if these laws were applied, all
works based on statistical reasoning would be accurate: "But even
when these conditions have been fulfilled correctly, only those works
seem justified which are based on very many individual cases and
which certify previously uncertain results and raise new points of
view."[38]

One group of medical researchers who took seriously advice like
Schweig's regarding the use of probability mathematics were epidemiologists associated with Max Josef von Pettenkofer (1818–1901).
Pettenkofer had entered the University of Munich in 1837, passed
with honors the state qualifying examinations for both pharmacist
and physician by June 1843, and spent the following summer at the
University of Giessen, where he became an enthusiastic disciple of
Liebig. In 1865 he was made ordinary professor of hygiene and
elected university rector at Munich. During an audience with the
young King Ludwig II, Pettenkofer promoted hygiene so effectively
that chairs were created at the universities of Würzburg and Er-
langen, and the subject was made compulsory in state medical examinations. In 1865, with Carl Voit and two associates, Pettenkofer
founded and for eighteen years coedited the *Zeitschrift für Biologie*,
which published many of his reports.

Pettenkofer's initial and continuing preoccupation with epidemiology stemmed mainly from his determination to explain recurrent
outbreaks of cholera. After his appointment as a member of a commission to review scientifically the 1854 epidemic in Bavaria, he formulated the hypothesis that the decisive factor in the genesis of cholera was the moisture content of the local soil, which was indicated by

the groundwater level. In direct opposition to the view that cholera was water-borne, as espoused by John Snow and William Budd in Great Britain, Pettenkofer argued that, when the water table fell, a larger soil volume became available for production of the toxic miasma.[39]

Pettenkofer's theory that there was a relationship between the level of groundwater and the spread of disease was extended to the study of typhoid fever by his student Ludwig Seidel (1821–1896) in an early article in the *Zeitschrift für Biologie*. In Seidel's analysis, considerations of mathematical probability were crucial. He tabulated the number of deaths from typhoid fever in Munich hospitals for each month in the year over the nine-year period 1856–65. Over the same time period, he displayed data on the level of groundwater in Munich. He found the average number of deaths per month to be 14.14 and the average level of groundwater to be 8.18 for all the data collected. Seidel then created a table in which he noted for each month whether the value for the number of deaths and the level of groundwater was more (designated by "+") or less (designated by "–") than the average. Although the two series of numbers were not entirely uniform, he found that there was a remarkable degree of correlation between them, that is, both tended to be either above the average or below it. Seidel concluded, "The supposition that chance would lead to such an uneven distribution between two equally possible cases ... is extremely improbable according to the well-known rules of probability theory to which I am referring here."[40] In particular Seidel observed that the limits by which an average would differ from an actual value could be determined by integrating over the quantity $(1/\sqrt{\pi})e^{-t^2}$, the equation for what in the nineteenth century was called the Gaussian distribution, or the astronomer's error law. On the basis of his calculations, Seidel concluded that the odds were 273 to 1 that the relationship between groundwater level and typhoid mortality "was not the work of chance, but founded in the fact that this conclusion is reached more readily than opposing theses in nature."[41]

Seidel's conclusions generated a response, also in the *Zeitschrift für Biologie*, from the Kiel psychiatrist Willers Jessen (b. 1824). Jessen explicitly attempted to compare Seidel's mathematical methods with those of Gavarret. The manner in which Jessen chose to frame his discussion clearly indicated that such a probabilistic approach was not the norm in the epidemiological thinking of the 1860s. He began by observing that Seidel's problems could also be handled by the methods of Gavarret—methods he characterized as "not superfluous." Then he reproduced some of the tables for the limits of oscilla-

tion derived from Gavarret's various formulas and noted in particular
that Gavarret had assumed a confidence level of 212:1 odds. Jessen
concluded:

> The results obtained here and from Seidel are in agreement, which is
> practical evidence that both methods are useful. Both also rely substan-
> tially on the same foundation, but Gavarret's method or at least its ap-
> plication will more easily allow itself to be presented popularly. Gavar-
> ret, himself, and after him, Fick[42] have already tried this successfully;
> perhaps now more than ever is the time to establish a general entry to
> analytical statistics. Wittstein rightly claimed at the meeting of scientists
> in Hanover that this science, although it currently barely exists, will in
> the future solve problems of which we as yet have no conception. The
> Munich examinations are proof of the truth of these words. Who would
> have thought a few years ago that one would be able mathematically to
> establish the relationship between a sickness and meteorological cir-
> cumstances? And yet this demonstration has really and undeniably been
> supplied through the common activity of the Munich researchers.[43]

Although Jessen was obviously enthusiastic about a mathematical
approach to medical statistics, he still chose to characterize such
methods as "novel." Even though German commentators on Gavarret displayed more mathematical sophistication than did their American, British, or French counterparts, such a probabilistic approach
was still marginal to mainstream German epidemiological thinking.

The novelty of probabilistic analysis was evident in one of the most
comprehensive treatises dealing with medical statistics from an epidemiological framework, the *Handbuch der medicinischen Statistik*
(1874) by Friedrich Oesterlen (1812–1877). Oesterlen's most notable
contributions were in the fields of physiology, hygiene, and statistics.
He had studied initially at Tübingen and then took a "scientific" trip
to Vienna, Würzburg, and Paris. He subsequently taught at Tübingen
from 1843 to 1846 before engaging in private practice in a variety of
German cities and eventually died in Stuttgart.[44]

Oesterlen was eager to demonstrate the various possible uses of
statistical methods in a medical context. In the introduction to his
treatise, he referred to the "great indifference of very many against
statistics ... even in these days."[45] Nevertheless, Oesterlen observed
that statistics were beginning to attain a degree of prevalence in medicine. He warned that one must be aware of the limitations of statistical reasoning: "Because if we want to use statistics in medicine, we
must know most of all what statistics can accomplish in a field like
ours, and what we can thus expect from it, and what we cannot."[46]
The main problems that Oesterlen cited with statistical reasoning

were the unreliability of the data and the variability of the phenomena under observation.[47]

Oesterlen focused on hygiene and questions of mortality and morbidity as the proper domains for statistical analysis in medicine. In the introduction to his work, he observed that the prevalence of mortality statistics in various regions and localities permitted a relatively self-sustaining discipline to come into being. Hygiene, in particular, served as a prominent context for medical statistics especially in the study of the comparative death rate for various diseases.[48]

The overall structure of Oesterlen's work (which numbered some 950 pages) further illustrated the largely demographic framework within which he conceived the statistical enterprise. He divided the work into three parts to provide an extensive catalog of the types of medical data about which statistics could be collected: the general statistical relations of populations, statistics of isolated diseases and different causes of death, and statistics of morbidity and the isolated causes of sickness.

Since Oesterlen was principally concerned with what statistical methods revealed about large-scale social patterns of the spread of disease, his discussion of the more mathematical aspects of statistical reasoning was brief. He commented on the novelty of such methods:

> Comparing masses with masses as in their statistical evaluation, is a product of modern times, especially in our field, and does not belong to the childhood period of the realm of research. Nevertheless, all too many researchers and physicians have shown how dangerous it is not to know the first rules of statistics, of probability.[49]

Oesterlen did not go into the theoretical details underlying the "rules of statistics and probability," that is, the fact that they were based on the assumption of a particular standard of 0.9953. Rather, he cited without proof the algebraic equation used by Poisson and Gavarret for the limits of oscillation given by $m/\mu \pm 2\sqrt{2mn/\mu^3}$ and then applied this formula to determine the possible oscillation annually for the number of male and female births based on data collected in France in 1825.

In the primarily epidemiological and demographic contexts of Pettenkofer's *Zeitschrift für Biologie* or Oesterlen's *Handbuch der medicinischen Statistik*, large quantities of data were available. The large number of observations required by Gavarret's assumption (following Poisson) of a standard of 0.9953 could be applied directly. In the context of clinical statistics, however, the requirement of several hundred observations to determine the efficacy of a therapy was more difficult to carry out. Consequently, most of the German commentators who applied Gavarret's methods to clinical results specifically engaged the issue of the high standard of reliability. They hoped that, by assuming a smaller value, they could make Gavarret's equations more readily applicable to the clinical statistics actually available and thereby render medical therapeutics more "scientific."

One of the first German commentators on Gavarret to raise this issue was Adolph Fick (1829–1901) in an appendix to the second edition of his textbook *Die Medicinische Physik* (1866). Fick belonged to a generation of German physiologists from the middle of the nineteenth century who were developing a science of physiology by physical methods and viewed life's processes from a mechanical point of view. Having entered the University of Marburg in 1847, he displayed an early aptitude for mathematics but decided to study medicine. He spent a good portion of 1850 at the University of Berlin, where he worked mostly in the clinics but was stimulated by Traube, Du Bois-Reymond, and Helmholtz.[50]

Fick's account of the role of probability theory in medical statistics was written in the style of a proselytizer; he wanted to show that a mathematical methodology should be adopted by those collecting clinical statistics. He wrote that such methods were justified provided "they were used in the right manner" and quoted Laplace that the theory was good sense reduced to a calculus.[51] Fick even criticized Valleix for failing to take Gavarret seriously:

> Gavarret has correctly challenged the conclusions which Louis drew from a series of observations on the effect of bloodletting . . . but Valleix in fact completely denies Gavarret's principles which he did not understand and defends Louis's conclusions. In that discussion much is said about the fact that in Louis's statistics, not only numbers played a role, but also precise observation. Of course, in this lies already a misunderstanding for, in the statistical method, precise observation has only the purpose of establishing what is to be counted; the process of proof lies purely in the numerical relationship. But Valleix believes he deals a mortal blow to the application of probability calculations by presenting Louis's proof for the efficacy of tartaric acid in pneumonia, which he thinks must be obvious to anyone, even though it contradicts the principles represented by Gavarret.[52]

Fick hoped that future commentators would not be as quick to dismiss Gavarret.

Fick explicitly addressed the issue of how the algebraic equations for the limits of oscillation were derived from assuming a standard of certainty of 0.9953 or 212:1. He observed that Poisson had assumed this value in his "famous work" on the application of probability to judicial opinion and that this value had also been used in Gavarret's text. Unlike other commentators (especially British and French medical writers), Fick realized the arbitrariness behind Poisson's initial assumption: "The selection of precisely this number [0.9953] does not at all have intrinsic reasons. It is chosen simply because the calculation becomes simpler and because it is a number close to unity (the symbol of certainty)."[53] Fick then reproduced one of Gavarret's algebraic expressions defining the proper limits of oscillation for such a standard of reliability: $m/r \pm \sqrt{8m(r-m)/r^3}$.

Fick did not choose to develop alternative equations for alternative confidence levels. Like all previous commentators, he applied Gavarret's formula directly without modification. He used a hypothetical case in which 180 deaths occurred from a disease observed in 900 patients, i.e., $m = 180$ and $r = 900$. By the formula given above, it follows that the average mortality would vary between approximately 16% and 24%.

Fick also showed how Gavarret's formula could be used to judge whether or not a new therapy was efficacious. If 700 cases of the disease used in the preceding example were treated by a new therapy with only 84 deaths, then the question arises whether this reduction in mortality was due to chance or the new therapy. Gavarret had declared that, in such cases, one should add the values under the square root for the limits of oscillation: $\sqrt{8m(r-m)/r^3 + 8m_1(r_1 - m_1)/r_1^3}$. In this case, $r = 900$, $m = 180$, $r_1 = 700$, and $m_1 = 84$. The value of this square root was then computed to be 0.0513. However, the difference in mortality rates for the two series was given by $m/r - m_1/r_1 = 0.20 - 0.12 = 0.08$. On the basis of these computations, Fick concluded, "This difference is greater than the computed size of the root (0.08 > 0.0513) and so we can bet more that 212 to 1 that the medicine has an effect on the sickness and a good one at that, since it has lessened the average mortality."[54] Even though Fick had raised the issue of the arbitrariness of Gavarret's high confidence level, he did not elaborate on how the formula would be changed if a smaller value were assumed.

This hope that Gavarret's formula could be made more directly applicable to clinical statistics by assuming a smaller standard was seized by subsequent German commentators. One such individual was the ophthalmologist Julius Hirschberg (1843–1925), whose educational background was strikingly similar to that of Fick. After graduating from a classical gymnasium in Potsdam in 1862, he studied medicine at Berlin. Among the teachers who influenced him were Helmholtz, Du Bois-Reymond, and Traube. He became a doctor on March 21, 1866, by writing a thesis on the treatment of placenta praevia under the direction of no less a figure than Rudolph Virchow. Subsequently Hirschberg attended lectures on higher mathematics and on optics and went to Paris, Vienna, Prague, and London to further his education. Through Virchow, Hirschberg was made the assistant to the ophthalmologist Albrecht von Graefe for the period 1866–68. In 1869 he was able to set up his own private clinic in Berlin as an eye specialist. In the following year, he was appointed *Privat-Dozent* at the University of Berlin, formally inaugurating his career in academic medicine (he was made ordinary professor in 1900).[55]

Drawing on his mathematical training and motivated by his desire to show by statistical methods the efficacy of some of the newer therapies advocated by von Graefe, Hirschberg produced a work in 1874 entitled *Die mathematischen Grundlagen der medicinischen Statistik, elementar Dargestellt.* Like other commentaries, it was designed to show how "novel" methods like those of a Gavarret could be put to use in the context of clinical statistics. After a passing reference to the debates in the Parisian Academies of Science and Medicine in the late 1830s, he observed that physicians had not been convinced of the utility of introducing probability into medicine because they believed "probability is not able to make science." Hirschberg mentioned, in contrast, that probabilistic reasoning is "frankly indispensable" in several areas of science and that it in no way "deserves the disdain with which several physicians look down on it." He cited the use of probability in astronomical observations as an instance of its deployment in science.[56] Hirschberg was trying to resuscitate probabilistic considerations as they applied to clinical statistics.

Hirschberg saw probability theory as the key to providing medical statistics with a scientific foundation. As he observed in the introduction to his work, "A really scientific explanation of medical statistics through probability calculations will dispose of the shameful judgments which one sees so often in today's medical literature (. . . Statistics can be made to prove anything—*Edinburgh Medical Journal*)."[57] He noted that the attempts to provide a probabilistic foundation for statistical reasoning had extended from the work of Jakob Bernoulli at the beginning of the eighteenth century to the 1840 text by Jules Gavarret. However, Gavarret's work had received so little attention from the medical profession that "it had to be discovered again so to speak by Professor A. Fick."[58] Hirschberg attributed this lack of interest in Gavarret to the aversion to mathematical discussion in general among physicians.

Even though Hirschberg criticized the medical profession for ignoring Gavarret, he worried that physicians might accept Gavarret's conclusions at face value. He noted that Gavarret's equations had been borrowed directly from the work of Poisson: "The proof of these questionable theses is however not at all self-evident, but rather quite complicated. Those who tend toward skepticism could question the correctness of Gavarret's discussion as much as those of any other medical hypothesis, such as homeopathy."[59] It was not so much Gavarret's text in particular, but rather the general introduction of probabilistic reasoning into the training of the physician, that would ensure advances in medicine; probability was another aspect of "scientific" training that the physician ought to possess. Hirschberg concluded, "If the prospective physician is to be a natural scientist, he must be particularly encouraged to study probability calculus."[60]

Hirschberg's principal concern was the number of observations required by Gavarret's assumption of 212:1 odds. He observed that smaller numbers of observations were not worthless, but their value "must be correctly judged." The level of probability that could count as sufficient had been arbitrarily chosen; Gavarret had adopted Poisson's standard because it had a "certain historical value." Hirschberg continued:

> Which businessman would not undertake a trade for which success he had reason to bet 212 to 1? The medical trade is in fact led by other motives than business speculation. But if absolute certainty is not to be calculated, one will nonetheless have to accept such a high level of probability as a substitute for certainty, just as judges and jurors do not abstain from giving a verdict based on such probability.[61]

Even though Hirschberg had cited the historical precedent of 212 to 1 odds, he believed actually that a lesser standard of reliability could be used—largely out of loyalty to his teacher von Graefe. He had collected statistics on procedures for removing cataracts—the method advocated by von Graefe and another method. For von Graefe's method, the average failure rate was found to be 5% with a limit of oscillation given by 2%, assuming Poisson's confidence level of 0.9953. The alternative method generated a failure rate of 3.3% with a possible variation of 2.9%. Even though there was a difference between the success rate of the two methods (5 − 3.3 = 1.7), the possible limits of oscillation for both series was measured to be 3%. Since the difference in success rate for the two methods was less than the possible variation, no definitive conclusion could be reached as to the relative merits of either approach. As Hirschberg observed, "From the above numbers, one is still also not able to conclude with the probability of 0.995 that v. Graefe's method is preferable."[62] Hirschberg proceeded to modify the various formulas for the limits of oscillation by selecting the lower standard of reliability of 1:11 or 0.916. From one series of data, he found the comparative success rate for the two approaches to differ by 2.2% with a possible variation of only 1.7%. Since the variation was less than the comparative difference, Hirschberg concluded, "By basing [the results] on the smaller probability of 0.916, the second series speaks in favor of v. Graefe's method."[63]

Hirschberg's account of the mathematical foundations of medical statistics was unique both in terms of the mathematical detail present and in terms of its sheer length of ninety-four pages. Most other accounts of Gavarret had been either smaller parts of larger works or journal articles. As an extended monograph, Hirschberg's work became itself a subject for review in the medical literature. These commentaries highlighted the lack of mathematical knowledge among the medical profession as a whole. As an anonymous review of Hirschberg's work in the *British Medical Journal* observed:

> We believe that the advantages of *statistics*, in other words, of long tables of cases, or their results crystallized into a formula, have been greatly overrated in medicine. At the same time, the advantages of the statistical method, and the clearness and definition given to our diagnosis and practice, and to its records, by the habit of mind engendered by mathematical studies can scarcely be overvalued.... One thing is quite evident, that such methods would give the deathblow to most of our modern systems of quackery.[64]

Although Hirschberg's text was clearly admired for its mathematical sophistication, it was still discussed as an innovative approach in the collection of clinical statistical rather than a standard research practice.

Other German university-based clinicians who commented on Gavarret addressed issues similar to those Hirschberg raised; they wanted to reduce the standard of certainty so that probabilistic reasoning could provide a "scientific" basis for medical therapy. One such commentator was Carl Liebermeister (1833–1901), who had studied medicine in Bonn and Würzburg before eventually teaching in university clinics at Basel and Tübingen.

In an 1877 article published in *Volkmanns Sammlung klinischer Vorträge*, Liebermeister addressed the issue of the excessive number of observations required by Gavarret's mathematical formulations. He noted that by the time the demands of probability theory were united with therapeutic statistics, the question of the degree of certainty was moot because the number of observations had become so large. Nonmathematicians required a formula that had more practical value. Liebermeister framed the issue as a disciplinary dispute between mathematicians and physicians:

> They always demand at least hundreds and often many thousands of individual observations. Now it is easy for the mathematician to say: you physicians must, if you want to draw certain conclusions, always work with large numbers; you must put together thousands or hundreds of thousands of observations. But that is usually not possible with therapeutic statistics. . . . Only in rare cases of therapeutic statistics can the conditions be fulfilled in the ways that mathematicians have demanded up to now.[65]

Liebermeister observed that the proportion would still tend toward certainty after a small as well as a large number of observations, but the degree of certainty would be less; he referred to Hirschberg's assumption that 1:11 odds were sufficient. Invoking a monetary metaphor, Liebermeister declared: "In the empirical foundations of therapy we are truly not so rich in gold coins that one could advise us to throw all the silver coins into the water! And a handful of silver coins is often worth more than a single piece of gold."[66] In an example he showed how even 5:1 odds could be assumed as an appropriate standard that would be useful in actual practice.

Shortly after Liebermeister's account appeared two more articles on the general relationship between probability theory and the determination of medical therapy by Friedrich Martius (1850–1923). Martius had studied under such pioneers of bacteriological research as F.A.J. Loeffler (1852–1915) and Emil von Behring (1854–1917) before eventually settling as a physician at a military hospital in Hanover.

Martius wrote his two articles on the relationship between probability and medical therapy out of a desire to "give foundations to medical science." These articles, published in 1878 and 1881, inaugurated his career in medical research, which would span approximately forty years and would result in approximately seventy publications. The two articles analyzed previous discussions of Gavarret; however, they were the only articles that Martius ever wrote on these subjects. He later pursued a career in experimental research working at the Institute of Du Bois-Reymond in Berlin, where he eventually met Helmholtz.[67]

Coming at the end of a forty-year commentary on the principles of Gavarret, Martius's analyses served as a nice capstone of nineteenth-century views of the role of probabilistic mathematics in helping to provide a "scientific" basis for medical therapy. In the first article Martius framed his analysis of scientific research in therapy in a historical context. He observed that a tension had always existed in medicine between the claims of rational dogmatists and empiricists. The former built abstract metaphysical systems and the latter placed undue emphasis on the mere collection of facts. German medicine had been influenced in the first half of the nineteenth century by metaphysical dogmatists as a result of *Naturphilosophie* but had now "sailed safely out of the stormy waters of dogmatism."[68] Since excessive rationalism and excessive empiricism were both unsatisfactory, Martius proposed to analyze theoretically the best method to follow in determining therapy. He chose the works of Gavarret and Poisson as providing appropriate tools for carrying out such a discussion.

In the work of Gavarret, Martius observed that four objectives had been addressed: to understand all the circumstances of the sickness (e.g., temperature, constitution of the patient, previous illness, etc.), the hygienic conditions preceding the sickness, the hygienic conditions during the treatment, and the sickness itself. Martius warned that the variable effects of therapies made the strict application of the rules laid down by Gavarret difficult in actual practice:

> The blame lies merely with the objects of research in scientific therapy, the analysis of which throughout is still not so well founded as to allow a strict application of the experimental method. Therefore one is hardly permitted, as Gavarret in his initial enthusiasm for his principle alleged, to speak of the numerical method with its conclusions of probability as the highest and most perfect level of research methods that are useful for therapy. It is and remains much more a makeshift necessity.[69]

Martius's view of probabilistic methods as merely a "makeshift necessity" derived from his laboratory training which had instilled the ideals of certain knowledge. He observed that therapeutic research would have to entail physiological deduction going "hand in hand" with exact, empirical, clinical induction. Martius summarized his view by quoting from the physiologist Karl Vogt, who had maintained:

> The practicing physician cannot be an experimenter. But what one can and must demand is that the physician during his studies be taught experimental methods, methods to discern truth from falsehood, so that he can form his own critical judgment about applied methods and

about received results in order to know what he should look for in the differences of available opinions, and which conclusions he should draw from his observations.[70]

Despite his skepticism regarding the ability of probability theory to render medical therapy scientific, Martius produced a more extensive analysis of the subject in his second article "to contribute still further to the clarification . . . [of] the logical foundations of statistics and probability."[71] The analysis was based on a short talk Martius gave at the military physician society in Berlin and was eventually published in 1881 in the *Archiv für pathologische Anatomie und Physiologie und für klinische Medicin* on the basis of Martius's personal visit to the journal's editor, Rudolph Virchow. Even forty years after the initial publication of the article, Martius still regarded it as the best thing he had ever written.[72]

Martius again attempted to put Gavarret's application of probability theory to medical statistics into a historical framework. He maintained that the first attempts to use a kind of "numerical method" in medical therapy had occurred nearly a half century previously and that "Louis and Gavarret were the ones in particular who, with the usual ardor of reformers, believed they should inaugurate a whole new era of scientific medicine. To be sure, their hopes and wishes were not fulfilled."[73] Most physicians continued to distrust the efficiency of the numerical method when combined with probability theory. Like Hirschberg, Martius attributed this distrust to the more general "mathematical unfitness" of the medical profession as a whole.[74]

Martius also commented on the standard that Gavarret had assumed in deriving his formulas. He observed that Gavarret's initial assumption of 212:1 odds had been adopted by "almost all theoreticians concerning medical statistics, such as Schweig, Oesterlen, Fick, [and] Hirschberg."[75] Martius then cited Fick's formulation of Gavarret's law: $m/r \pm \sqrt{8m(r-m)/r^3}$.

Martius particularly singled out Liebermeister's attempt to make Gavarret's formulas more "practical" by assuming a smaller standard than 212:1. By highlighting the arbitrariness behind the confidence level assumed, Liebermeister's approach could lead to contradictory therapeutic opinions regarding the probable value to assign to isolated mean values. According to Martius:

Since there remains a subjective preference whether one uses the greater or the lesser probability in the comparison of therapeutic averages, it can easily happen that after the computation of one formula, the superiority of one healing method over another seems proven, while the use of the other formula with the same data can make the positive outcome of the first seem merely coincidental! And this privileged arbitrariness is supposed to serve to "establish according to principles of physics the effect of different healing methods on a sickness!"[76]

Since Liebermeister had implicitly reintroduced the arbitrariness into medical therapy which the invocation of mathematical reasoning had been designed to eliminate, Martius concluded that "Liebermeister's formula, although also practically useful, is, however, more unscientific than that of Poisson."[77]

Martius viewed with detachment such debates over how to render medical therapy "scientific" by applying mathematical probability. As one trained in laboratory methods, he retained the faith that the basis for science lay in experimentation rather than mere observation and the collection of numerical data—regardless of how those data were mathematically manipulated. Martius concluded his analysis with the observation, "The real progress of medicine, that which sheds light on causal relationships of phenomena, lies in experimental induction, not in the numerical method."[78]

With Martius's articles, the forty-year commentary on Gavarret's probabilistic views came to an end. As an intellectually hybrid discourse combining statistical thinking with probability mathematics, it failed to generate a receptive audience within the various nineteenth-century medical communities. It was always discussed as a "novel" method that might be applied to some specific problem at hand (such as the relationship between groundwater level and typhoid fever or the success of cataract operations); it was never discussed as the "standard" method of adjudicating claims in therapeutic disputes. For physicians such as William Guy or Louis René Villermé whose primary professional home was the nineteenth-century statistical society, the social changes illustrated by aggregative thinking were more pressing than Gavarret's mathematical considerations. In contrast, for physicians such as Hirschberg and Liebermeister whose primary professional home was the university-based clinic, Gavarret's text was theoretically useful in providing a "scientific" basis for therapy; however, the number of observations required was impractical. These writers attempted to reduce the standard of 212:1 odds that Gavarret had borrowed directly from Poisson. As Martius's analysis pointed out, however, the arbitrariness behind the assumed standard undermined the broader attempt to use Gavarret's work as a tool to make therapy "scientific." In the realm of medical therapy, probability mathematics and clinical statistics remained largely separate spheres of intellectual inquiry throughout the nineteenth century.[79]

Even though Gavarret's specifically probabilistic approach to medical statistics was far too mathematically technical to generate sustained support in the nineteenth century, the more general issue posed by the 1837 debate of the Academy of Medicine—namely what constituted the "science" of medicine continued to be hotly contested. With the advent of laboratory methods, the research physiologist emerged as a strong contender to wear the mantle of the science of medicine. In order to claim the title successfully, however, the physiologist would have to come to terms with Louis's argument that qroup quantification was the key to making medicine into a science.

THE LEGACY OF LOUIS AND THE RISE OF PHYSIOLOGY: CONTRASTING VISIONS OF MEDICAL "OBJECTIVITY"

As OBSERVED in the earlier discussion of Louis, his historical signifi-cance had more to do with his general claim that the clinical physi-cian should aspire to become a scientist than with his specific "nu-merical method" of direct comparison between alternative therapies. This distinction is key to understanding Louis's bifurcated legacy: his rhetoric regarding medicine's scientific foundation survived even though it was eventually appropriated by research physiologists—who explicitly rejected Louis's clinical orientation; his method, by contrast, eventually became so standard in determining the efficacy of new therapies that its association with Louis's "scientific" vision was lost. This chapter first discusses Louis's immediate leg-acy as president of the *Société Médicale d'Observation*, examines famous nineteenth-century therapeutic innovations based on nu-merical comparison (such as Lister's demonstration of the efficacy of antiseptic surgery), then explicates the reasoning behind Claude Bernard's desire for medicine to become an experimental laboratory-based science, and concludes by demonstrating how medical obser-vation came to be seen as more "objective" during the course of the nineteenth century even though problems of medical inference (i.e., when was a therapy truly efficacious?) still depended on the in-formed professional judgment of the individual clinical or physiolog-ical researcher.

Louis had an immediate, though short-term, impact on the Pari-sian medical scene when some of his followers founded the *Société Médicale d'Observation* in 1832. This society was committed both to Louis's general vision of medicine as an empirical science and to the specific conclusions that could be derived by following the numerical method. Louis was named perpetual president with Chomel and An-dral as honorary presidents. Until his death in 1855, Valleix was the vice president of the society. The society produced three sets of memoirs that derived conclusions by using the numerical method; these memoirs appeared in 1837, 1844, and 1856 respectively. Ini-tially, 46 individuals were associated with the society (including a significant number of Louis's American students[1]). When the final memoir appeared in 1856, a total of 149 individuals had been associated with the society at one time or another.[2] By this time, however, Louis's influence on Parisian medicine had already begun to wane; he had retired from public life following the premature death of his only son Armand in 1854.[3]

Even though the *Société Médicale d'Observation* was short-lived, the publications from the three sets of memoirs provide clear insight into Louis's philosophy that medicine should aspire to becoming an empirical science. Speaking *ex cathedra* as president of the *Société Médicale d'Observation*, Louis declared in 1837 that "medicine is a science of observation, that it is entirely."[4] Ultimately, Louis differed from his critics more in his orientation toward aggregative thinking than in his actual medical outlook. Like his critics, he paid attention to such distinguishing factors as age, sex, and patient history in collecting his results. He even warned that "it is not sufficient to count in order to know the relation between the facts; and in employing the numerical method without discernment, either one can arrive at absurd results, or results that experience does not verify."[5] Louis did not deny the individuality of the patient, as his critics had charged; rather, he maintained that further "scientific" insight could only be attained by focusing on similarities that became evident through viewing a population as a whole. In his opinion, "if there is a method to collect the experience of centuries, this can only be by employing the numerical method."[6]

Louis's followers also emphasized that, although they recognized the individuality of the patient, the clinical physician could acquire a scientific standing only by following an approach predicated on aggregative thinking. Valleix explicitly framed the issue in these terms in his analysis of the Academy of Medicine debate of 1837. He recounted the arguments against enumeration, namely that each individual malady was a distinct entity and that excessive concern with empirical fact gathering would arrest the "spirit of genius" and deprive science of good discoveries. In response, Valleix argued that all physicians drew inferences from past experience; one had to recognize similarities in order to diagnose disease. He maintained, "Everyone searches, the members of the society as all other physicians; the difference consists in the manner in which we search, which constitutes the analytical and numerical method, the only one which is able to lead to truth."[7] Valleix declared that even G. B. Morgagni (1682–1771), who had criticized enumeration in the eighteenth century, would give greater credence to evidence based on a larger number of observations and that all instances of genius in medicine had to be based on evidence that could be proved true only after a sufficient number of cases.[8] Although the use of enumeration might lead to dull reading, it permitted medicine to acquire scientific standing: "The chemist, the physicist, the mathematician, when they give the response to a scientific question, the solution to a problem, do they propose to amuse or to instruct? If, by a sad exception, medicine abstains by sacrificing the foundation to the form, it will lead to the despair of our science."[9]

Valleix gave a concrete example of how such a numerical method could be applied in an article that he wrote in the second volume of the *Mémoires de la Société d'Observation*. He measured the pulse in a group of infants for several days producing a total of 567 observations. From these observations, Valleix arrived at the following general conclusions: the mean pulse rate was 87 beats per minute when the infants were asleep and between 90 and 100 when they were awake; elevation of temperature led to a rise in pulse rate; boys have a more frequent pulse than girls do.[10] For Valleix, such conclusions possessed scientific validity.

Immediate commentators on Louis (both supporters and opponents) accepted Louis's assertion that enumeration constituted science. They might reject Louis's specific claim that medicine could become an empirical science; however, they never rejected the broader claim that enumeration of empirical facts was a key facet of scientific reasoning. The Harvard-educated American physician Austin Flint (1812–1886), who engaged in polemics over the proper type of "scientific" training for the physician throughout his entire career,[11] declared in an early article supportive of Louis's numerical method that "Baconian philosophy, properly applied, is fully adequate to the development of truth in all departments of knowledge."[12] Medicine was a branch of knowledge in a state of "infancy contrasted with manhood" when compared with other fields such as chemistry, astronomy, and physics. It was more on a par with political economy and meteorology in terms of the certainty of its methods. In 1843 the Irish physicians William (1794–1848) and Daniel Griffin (?–?) made similar arguments in a treatise on medical practice. Astronomy, chemistry, and optics had moved from positions of uncertain to certain knowledge while medicine had failed to do the same.[13] They prophesied: "The 'numerical method' then, is the only one medicine can look to with any hope of attaining certainty as a science. ... It cannot be doubted that the science of medicine is on the eve of a great and mighty revolution."[14]

Even as outspoken an opponent of Louis as François Double conceded the point that enumeration constituted science. He just believed that the nature of medical observation was too imprecise for counting to provide much insight. Thus, the attempt to make medicine into a science was a fundamentally misguided enterprise from the outset. As Double framed the issue in an introduction that he wrote in 1842, the year of his death, the attempt to make the numerical method anything more than an auxiliary or complement to experience was a "dangerous usurpation." Numerical calculation applied rigorously "will distort science and give in its place an appearance of certitude that is not permitted; applied as the way of mobilizing facts, it loses a good part of its value."[15] Double reaffirmed that medicine was fundamentally different from the physical sciences:

> Why is the science of man still so far from the physical sciences on which proposals have been made to model it without ceasing? . . . It is because there is no similarity. Here [in the physical sciences] the observation produced is always the same; there [in the human sciences] it varies without ceasing. It is this prodigious variation that created difficulties in the physiological sciences.[16]

With the death of Double in 1842 and Louis's retirement from the medical scene by the mid 1850s, a process began whereby Louis's specific method of taking averages was gradually disassociated from his more general claim that this method constituted scientific reasoning. Some medical commentators from the 1850s began to argue that the compilation of numerical or statistical results might provide some useful insights about therapy; however, these results did not possess the authoritative status of "science." One of the first such commentators was Friedrich Oesterlen (1812–1877), who would later provide one of the most extensive compilations of medical statistics produced in the nineteenth century. Oesterlen displayed skepticism toward the scientific standing of statistical comparisons in his 1852 work *Medicinische Logik*, a general treatise on the philosophy of science with a particular focus on medical issues. On the question of the comparative merits of the numerical method or statistics, Oesterlen claimed that such conclusions were nothing more than facts founded on experience. As such, they were valuable for providing useful data. However, "scientific" results could be obtained only after the determination of the causal connections that governed these relations. On the more specific question of the usefulness of the calculus of probabilities, Oesterlen maintained with Louis's critics that the uniqueness of the individual patient precluded its widespread applicability:

> We cannot speak of a calculation of probabilities in the strict sense of the term; for the great complexity and constant variation of all the cir-

cumstances render the safe application of general statistical results to an individual case impossible, and thus, make every attempt to submit the latter to exact computation a mere illusion.[17]

Oesterlen's views were echoed three years later by William Pulteney Alison, one of the leading figures in the Edinburgh medical world. In a report in the *BAAS* (British Association for the Advancement of Science), Alison argued that statistical results may often prove useful in determining therapy even if the "scientific" reasons underlying the therapy's efficacy remain unknown:

> The object of this paper was to show that, notwithstanding the plausible objections often made to statistical inquiries, as being applicable to the support of so many principles, as to give little real support to any, there are various questions in medical science, of the utmost practical importance, which admit of a perfectly satisfactory solution in this way, and [italics in original] *in no other*, because the present state of *science* [my emphasis] does not enable us, nor afford any prospect of our being soon enabled to understand the intimate nature either of diseased actions, or of the powers by which they may be excited or counteracted; in many instances, when, by simply empirical observation, and comparison of numbers, i.e., by evidence truly statistical, although often not formally expressed as such, principles may be established which are already amply sufficient for practical application of the highest importance.[18]

Alison cited smallpox vaccination and I. P. Semmelweis's conclusion that the decaying matter from cadavers contributed to puerperal fever as therapeutic innovations that had been established solely by statistical methods.[19]

By the early 1860s, this view that statistical comparison or the numerical method was simply a heuristic aid to medical practice rather than the basis for medicine's scientific foundation was being voiced even by such a prominent member of the *Société Médicale d'Observation* as Armand Trousseau (1801–1867). After receiving his initial training in Tours, the city of his birth, Trousseau had came to Paris in 1825 to make a career in medicine. He soon established a professional connection with Louis because they were both members of a commission sent to study the effects of an epidemic of yellow fever in Gibraltar in 1828. The results of the commission were published in a two-volume study in 1830 and Louis summarized the findings of the commission in the second volume of the *Mémoires de la Société d'Observation*. He claimed to have established rigorously by comparative statistics that the nature of the fever was sporadic in Gibraltar.[20] Trousseau's bona fides as a disciple of Louis was unquestioned.

However, in a series of lectures that he delivered at the Hôtel-Dieu between the years 1861 and 1864, Trousseau explicitly rejected Louis's claim that the clinical physician should become a kind of empirical scientist. Speaking in his official capacity as a professor of clinical medicine, he emphasized the primacy of clinical training. The pursuit of any type of scientific knowledge was subordinate to learning how to diagnose and treat the sick. The numerical method was not rejected because it failed to offer insight about therapeutics, but because such invocations of statistical reasoning were used to justify the claim that medicine was a science rather than an art:

> I do not reproach the numerical method because it numerates, but I reproach it because it only numerates: in a word, because it depends, like the mathematician, upon an absolutely exact result. . . . Those who admire the numerical method, applaud consequences which I deplore; they do not wish for the intervention of intellect; I do—I wish to see intellect exercising itself in all its power.
>
> I am anxious to make myself clearly understood; I employ statistics, I even employ, if you like the numerical method, provided it be only regarded as a means sometimes preparatory, and most frequently complementary; but I spurn it with all my energy when it pretends to be a method complete in itself, and capable of conducting us, as a matter of necessity, to truth.[21]

Trousseau's concerns with the numerical method were being echoed at the same time in America. As David W. Cheever observed in his Boylston Prize Essay on "The Value and the Fallacy of Statistics in the Observation of Disease":

> The advocate of the numerical method is sometimes as one-sided as the specialist. He is ready to forget that figures are not brains, tables not perceptions, and that recorded observations do not give the power of observing. The statistician is but too often as fallacious and extravagant in his conclusions as those who rely exclusively on physical signs; both equally overlook the rational part of medicine. . . . "Of all dangers, a fallacious certainty is the greatest. A simple process of verification *a posteriori,* like the numerical method, never can be elevated to the dignity of a system, since it will be eternally true in medicine, that the problem is the individual." We know that this method must be still more incompetent for the treatment of disease.[22]

Even the leading advocates for statistical methods in the nineteenth century, Quetelet and his British popularizer, the astronomer and natural historian John Herschel (1792–1871), distanced themselves from Louis's claim that the use of statistical comparison conferred scientific standing on medicine. To a significant extent, this derived from their own professional vision of the function of statistics as a scientific enterprise; they thought it should be linked more to the premier exact sciences of mathematics and astronomy, in which they had been trained, rather than the imprecise art of medicine. On the question of the value of medical statistics, Quetelet declared, "Nothing has been more strongly contested . . . and from the manner in which they are applied it should be so."[23] He duly noted the objection to medical statistics derived from the individuality of the patient. Also, he cited "the abuse of statistics" when doctors slavishly treated all patients the same way without considering their constitution, their age, or their sex.[24] Nevertheless, he maintained that there was often sufficient similarity between cases for physicians to draw on analogies from past experience. As soon as such similar observations were collected "so as to render them comparable and to draw inferences from them, statistics have been formed."[25] He cited the fact that the pulse is quicker in an old man than a man in his prime as a result successfully established by statistics.

Quetelet, however, was not noticeably interested in the question that had most engaged the clinicians, namely therapeutics. He maintained that "different kinds of treatment have less influence on mortality than is generally supposed."[26] Mortality depended more on the particular organization of the hospital than on the actual treatment employed. Quetelet's interests remained focused more on broadly social questions like hospital administration than on the particular issue of what treatment should be given; he attempted to stay above the fray in what was essentially an intraclinical dispute.

This same ambivalence toward the role of medical statistics in advancing medical science was evident in John Herschel's famous discussion of Quetelet in the 1850 *Edinburgh Review*. Herschel included medical therapy in an extended list of the subjects to which statistical reasoning might offer insight:

> Men began to hear with surprise, not unmingled with some vague hope of ultimate benefit, that not only births, deaths, and marriages, but the decisions of tribunals, the results of popular elections, the influence of punishments in checking crime—the comparative value of medical remedies, and different modes of treatment of diseases . . . might come to be surveyed with that lynx-eyed scrutiny of a dispassionate analysis, which, if not at once leading to the discovery of positive truth, would at least secure the detection and proscription of many mischievous and besetting fallacies.[27]

Nevertheless, Herschel advocated caution in applying statistical reasoning to medical issues: "The comparison of multitude with multitude, the destruction of errors by mutual collision, and the slow emergence of truth from the conflict by its outstanding vitality, belong to a maturer age of science than that in which medicine had its origin or attained its present importance."[28] Although Herschel acknowledged the work of Louis and noted the value of statistics in the classification of disease, he maintained that professional standards (in particular the moral imperative to heal the sick) militated against the use of statistical methods in medicine: "It would require a physician of no common forbearance to abstain in fifty out of each hundred cases from the use of *all active medicines*—and of no common candour and defiance of professional censure to declare that he had done so, and to put on record the *failures* of this line of treatment."[29] Even these leading advocates of statistical thinking in the nineteenth century questioned the use of the numerical method, largely because it undermined the physician's professional obligation to heal the sick individual.

By the last third of the nineteenth century, the divorce between Louis's numerical method and the appeal to a rhetoric of science had become complete. This disjunction was illustrated very concretely in the arguments of Joseph Lister in his pioneering work with antiseptic surgery. In 1870 he published a series of articles in the *Lancet* showing how sterilization reduced the mortality rate for pyaemia, erysipelas, and gangrene infections following surgery. He cited statistics comparing the average mortality rate for all surgical procedures performed on his ward at the University of Edinburgh in the years 1864–66 (before antiseptic methods were introduced) with the average mortality rate for all surgical procedures performed in the three-year period 1867–69 (after the introduction of antiseptic methods). He found that, before antiseptic methods were introduced, there were 16 deaths in 35 cases or one death in every 2.5 cases. After antiseptic methods were introduced, by contrast, there were only 6 deaths in 40 cases or 1 death in every 6.66 cases. Lister concluded that "these numbers are, no doubt, too small for a satisfactory statistical comparison; but when the details are considered, they are highly valuable with reference to the question we are considering."[30] Although he did not attribute it to Louis, Lister was applying Louis's numerical method.

However, Louis's broader claim that such use of numerical comparison conferred scientific validity on therapeutic results was explicitly rejected by Lister. Louis's method had been empirical; it established the fact of therapeutic efficacy without worrying about the causal factors that made the therapy efficacious. This empiricism was unacceptable to Lister, who claimed that the germ theory of disease as pioneered by Pasteur accounted for the reduction in mortality following sterilization. The statistical results simply provided further evidence to bolster the more scientifically fundamental germ theory of disease. For Lister, this was the reason that several of his British surgical contemporaries did not generate statistical results that were as dramatic as his. They tried blindly to follow his antiseptic procedure; however, they did not accept the germ theory on which his methods were ultimately predicated. As Lister observed:

> Want of success in many quarters has not arisen from any unwillingness to try a new mode of practice. On the contrary, the publication of my first papers was followed by a very general employment of the material which I happened to select for carrying out the treatment, and which, unfortunately for the principle involved, was then little known in British surgery, so that the striking results which were recorded were too often attributed to some specific virtue in the agent. The antiseptic system does not owe its efficacy to any such cause, nor can it be taught by any rule of thumb. One rule, indeed, there is of universal application—namely, this: *whatever be the antiseptic means employed* (and they may be various), *use them so as to render impossible the existence of a living septic organism in the part concerned.* But the carrying out of this rule implies a conviction of the truth of the germ theory of putrefaction, which, unfortunately, is in this country the subject of doubts such as I confess surprise me, considering the character of the evidence which has been adduced in support of it. Yet, without this guiding principle, many parts of the treatment would be unmeaning; and the surgeon, even if he should attempt the servile imitation of a practice which he did not understand, would be constantly liable to deviate from the proper course in some apparently trivial but essential detail, and then, ignorant of his own mistake, would attribute the bad result to imperfection of the method. For my own part, I find that, in order to approach more and more to uniform success, it is necessary to act ever more strictly in accordance with the dictates of the germ theory. Failure on the part of those who doubt or disbelieve it is therefore only what I should expect.[31]

With Lister, the dissassociation between Louis's method of numerical comparison and the claim that such a method provided a scientific foundation for medicine had become complete. The way was open for the experimental physiologist to enunciate an alternative vision of the science of medicine based on the techniques of laboratory experimentation.

This vision was spelled out in abstract and philosophical terms by the founder of nineteenth-century positivism, Auguste Comte, in his *Cours de philosophie positive.* Strongly influenced by the physiologist François-Joseph-Victor Broussais (who had motivated Louis's analysis of bloodletting), Comte believed that mere empiricism (as practiced by Louis) was a profound degeneration of the medical art:[32]

> Pure direct experimentation restrained between suitable limits can be very important for medicine, as for physiology itself: but it is precisely by the strict condition of never being simply empirical, and by always attaching itself ... to a systematic assembly of corresponding positive doctrines. It spite of the imposing aspect in the form of exactitude, it will be difficult to conceive, in therapy, of a judgment more superficial and more uncertain than that which would reside uniquely on this false computation of cases fatal or favorable, without speaking of pernicious practical consequences ... The geometers have sometimes honored a profoundly irrational aberration, in their childish efforts to determine, by their illustrious theory of chances, the number of cases appropriate to justify these statistical indications.[33]

Comte made his arguments against basing medical science on statistics at the abstract level of philosophical and social theory; the individual who would take up these arguments as a rallying cry for changes in institutional practice was Claude Bernard (1813–1878) in his medical classic *Introduction à l'étude de la médecine expérimentale* (1865) as well as his posthumously published *Principes de médecine expérimentale* (1947). Bernard's analysis came after he had already established his scientific reputation, and it was couched in terms of an act of disciplinary justification. Although he had begun his medical career as a hospital intern in 1839, his exposure to the teaching of François Magendie at the Collège de France convinced him that the "science" of medicine resided in experimental physiology.[34] It should be practiced as a discipline autonomous from medical practice and carried out in laboratories where original research could be conducted and future practitioners trained. To base medicine on statistics, as Louis had done in the clinic, would be to assume that medicine was merely a passive observational science rather than an interventionist experimental one. Such an experimental approach would guarantee secure knowledge by dominating the totality of the vital conditions that influenced a given physiological event or process. Those with appropriate laboratory training and facilities would no longer need to be merely empirical and statistical.[35]

Bernard's view that experimental medicine had to become a laboratory science reflected not only his personal desire for more adequate laboratory facilities, but also his belief that French medicine and science had gone into eclipse vis-à-vis the emerging German research establishment. French medical training was still centered in hospitals as it had been some thirty years earlier when Louis's numerical method had been debated. Both academic career advancement and the training of students centered more on clinical skill than on the ability to conduct original research in a laboratory. To Bernard this was anathema: "I consider hospitals only as the entrance to scientific medicine; they are the first field of observation which a physician enters; but the true sanctuary of medical science is a laboratory; only there can he seek explanations of life in the normal and pathological states by means of experimental analysis."[36] In addition, Bernard noted how Germany had surpassed France with its well-endowed biological research institutes:

> The scientific impulse, started in France, spread through Europe, and little by little the analytic experimental method entered the realm of biological science as a general method of investigation. But this method was perfected more, and it brought forth more fruit in countries where conditions for its development were more favorable. Throughout Germany to-day there are laboratories, called physiological institutes, which are admirably endowed and organized for the experimental study of vital phenomena . . . Scientific production is naturally in proportion to the means of cultivation which a science possesses; there is nothing astonishing, then, in the fact that Germany, where the means of cultivating the physiological sciences are most liberally installed, is distancing other countries in the quantity of its scientific production.[37]

Finally, with regard to the tendency in French scientific and medical education to emphasize lecturing rather than the acquisition of experimental skill, Bernard lamented:

> By pointing out, from a professional chair, the results as well as the methods of science, a teacher may form the minds of his hearers and make them apt in learning and choosing their own direction; but he can never make them men of science. The laboratory is the real nursery of true experimental scientists, i.e., those who create the science that others afterwards popularize. Now if we want much fruit, we must first care for our nurseries of fruit trees.[38]

As a result of his laboratory-based orientation, Bernard shared with Louis's critics a concern with the individual living organism; however, he repudiated their professional vision of medicine as an "art." For Bernard, medicine should advance beyond being a mere art, or an empirical science, to become an experimental science. Bernard's rejection of both Louis and his opponents (for different reasons) was nicely encapsulated in his view of Risueño d'Amador. He observed,

"The critique is good [the rejection of statistics], but the conclusion [medicine is an art] is false."[39]

> It is necessary to abandon all these pretensions that the physician has been an artist. These are false ideas that are good only to encourage, as we have said, laziness, ignorance, and charlatanism. Medicine is a science and not an art. The physician should aspire to become a scientist; it is only from ignorance . . . that one must be resigned to being empirical in a transitory manner.[40]

Likewise, Bernard criticized the claim that medicine was an empirical science as Louis had maintained. Medicine (in good Comtean positivistic fashion) had to adopt the method of scientific experimentation to advance beyond the stage of "a conjectural science based on statistics." For the laws of medicine to become scientific, they could "be based only on certainty, on absolute determinism, not on probability."[41] With regard to Louis's method of taking averages, Bernard wrote:

> Let us assume that a physician collects a great many individual observations of a disease and that he makes an average description of symptoms observed in the individual cases; he will thus have a description that will never be matched in nature. So in physiology, we must never make average descriptions of experiments, because the true relations of phenomena disappear in the average; when dealing with complex and variable experiments, we must study their various circumstances, and then present our most perfect experiment as a type, which, however, still stands for true facts. In the cases just considered, averages must therefore be rejected, because they confuse, while aiming to unify, and distort while aiming to simplify.[42]

Bernard demonstrated concretely his rejection of aggregative thinking in a debate with Charles Viollette, a professor of chemistry from the faculty of sciences at Lille in 1875. On the basis of a statistical comparison of the proportion of sugar content in the leaves of beets, Viollette concluded that the sugar in beets was produced in the leaves rather than the roots because there was a diminution in sugar content when the leaves were thinned. The average sugar content in 37 shed beets examined was 10.54% while the average content for the 40 beets not shed was 13.11%. Bernard declared that such a method of statistical comparison proved nothing:

> Empiricism precedes science. It unites groups of very complicated facts so they can be sufficiently analyzed, it generalizes results by the methods of statistics. However, statistical methods give us only the state of

things, they explain to us nothing; they can be used no doubt and receive applications; but the imprints rest always on a certain quantity
unknown and undetermined; they can supply us only conjectures,
probabilities; we can draw no certitude for the particular case. Experimental science determines by analysis . . . the precise and simple conditions of a phenomenon, [and] gives to us the explanation and the reason. It is the expression of scientific determinism and permits neither
exception, nor uncertainty.[43]

The contrast between Louis's and Bernard's view of science can be
seen as a function of a much more fundamental disagreement; they
had contrasting views of what constituted "objective" knowledge. For
Louis, the idiosyncratic individual (which his clinical opponents admired) must be rejected for the "objective" knowledge that emerged
at the level of the population. For Bernard, any conclusions based on
population thinking must be rejected because such results constituted merely statistical or probabilistic knowledge; they did not possess the "objective" determinism of conclusions based on experimental investigation of the individual living organism.

Louis's and Bernard's differing intellectual visions of objectivity
manifested themselves concretely at the level of research practice.
For Louis, the proper professional behavior for the medical researcher was to diagnose and classify diseases in the clinic by observing pathological anatomical conditions in patients. "Science"
emerged when these observations were numerically collected and
represented in tabular form. For Bernard, the proper professional
behavior for the medical researcher was to analyze an individual organism's physiological functions by performing experiments in a laboratory. "Science" emerged when the researcher had a completely
deterministic understanding of the functioning of the organism.

In contrast to both of these views of objectivity, there emerged a
third view among some clinicians, that precision measuring instruments would be the key to providing "objective" medical data. Instruments such as the thermometer were introduced which trans-
lated the physiological action of the body into quantitative form.
Such evidence was seen as so unambiguous that it contributed to
what Zeno Swijtink has called the "objectification of observation" in
the nineteenth century.[44]

Such a commitment to precision measuring instruments on the
part of some clinicians caused them to adopt some of the attitudes of
both Louis and Bernard. By continuing to see the clinic rather than
the laboratory as the institutional locus for research, these physicians

employed the aggregative or group thinking that Louis used. However, by arguing that their professional status derived from their use of precision instruments to measure physiological processes,[45] they were allying themselves with the view that the "science" of medicine resided in studying the individual physiological organism in the manner of Claude Bernard.

One of the earliest advocates for the view that precision measuring instruments would confer "objectivity" on medical observation was the German clinician Carl Wunderlich. As observed earlier, Wunderlich had begun his medical career by helping to found the *Archiv für physiologische Heilkunde*. In the preface to the first volume of this journal, Wunderlich had declared that the "science" of medicine resided in the understanding of individual physiology and he repudiated a statistical or numerical approach as being merely "empirical" rather than "scientific." Although Wunderlich was a clinician rather than a physiological researcher, he and Claude Bernard were intellectually kindred spirits.

The particular physiological process that Wunderlich chose to study was temperature variation. Earlier German physicians had already played a leading role in pioneering medical thermometry. In 1850–51, F.W.F. von Barensprung and Ludwig Traube began to use temperature signs as basic data to diagnose diseases, predict their course, and determine therapy. It was Traube who persuaded Wunderlich to take up the subject of thermometry.[46]

Wunderlich explicitly engaged the issue of the relationship between the use of the thermometer and "objective" observation in his landmark study *On the Temperature in Diseases: A Manual of Medical Thermometry* (1868) based on the observation of over twenty-five thousand patients over a twenty-year period. He maintained:

> The tendency of modern medicine to set the highest value for diagnostic and prognostic purposes, upon *objective* symptoms, and amongst these upon what are called *physical signs*, is undoubtedly a step in the right direction.
>
> Now, the temperature of a sick patient is both an "objective" and "physical" symptom, and the use of the thermometer must be classed with the "physical diagnostic" methods of percussion, auscultation, etc.; and whatever may be claimed for these as regards their significance and practical value may be claimed for thermometry with equal justice.
>
> Thermometry, however, has this advantage over all these applications of acoustics, an advantage of almost priceless value, inasmuch as it gives results which can be *measured*, signs that can be *expressed in*

numbers, and offers materials for diagnosis which are incontestable and indubitable, which are independent of the opinion or the amount of practice or the sagacity of the observer—in a word, materials which are physically accurate. Amongst all the phenomena of disease there is scarcely another which admits of such accuracy or is so reliable as the temperature[47]

The greatest gain that could be obtained from the study of thermometric observations was that "the alterations of temperature in disease are subject to fixed laws."[48]

Even though Wunderlich had clearly employed aggregative thinking in his attempt to determine the "fixed laws" of temperature in disease, he explicitly eschewed the method of taking averages (as practiced by Louis). On the calculation of a kind of arithmetical mean value for a given disease as experienced by a particular population, he observed, "A mere statistical estimation of the curves in the gross mass must obliterate all the peculiarities of the course followed by the temperature, and a mere numerical treatment of the numbers of different cases can only afford trustworthy data for the answer of certain definite questions."[49] Rather than studying *numbers*, Wunderlich advocated a careful analysis of the *form* of the curves generated by the measurement of changes in temperature during the course of a disease. From such detailed comparisons, he envisaged the possibility of creating "a sort of model curve, which may approximatively express the peculiarities of single cases."[50] Wunderlich was well aware, however, that any such "model curve" could err in actual medical practice:

> [I]t is only the copious stores of material at my command, and constantly repeated proofs of the correctness of my principles which allow me to hope that they do not caricature or contradict nature. Although I do not arrogate to myself the right to declare them to be the laws of pathological action, I still believe they may serve as a very useful clue to those who interest themselves in the thermometry of disease.[51]

Although Wunderlich clearly used quantification to formulate his "laws of pathological action," he did not believe that medical inference could likewise be reduced to formal rules; that is, the meaning of the "objective," instrumentally determined data still required the skilled interpretation of the clinical physician.

Even though German clinicians had taken the lead in pioneering medical thermometry, early supporters in other countries echoed the view that such precision instruments provided "objective" data. One particularly prominent advocate was John S. Billings, surgeon of the U.S. Army, who observed in a lecture at an international medical congress:

> The word-pictures of disease, traced by Hippocrates and Sydenham, or even those of Graves and Trousseau, interesting and valuable as they are, are not comparable with the records upon which the skilled clinical teacher of the present day relies. ... The temperature chart has done away with the errors which necessarily follow attempts to compare the memory of sensations perceived last week with the sensations of today.[52]

Despite the objectivity precision instruments give to the data, Billings, like Wunderlich, continued to emphasize the uncertainty of medical inference:

> In medicine, as in social science, we must depend for many facts upon the observation of conditions which occur very rarely, and which cannot be repeated at pleasure. ... A science of medicine, like other sciences, must depend upon the classification of facts, upon the comparison of cases alike in many respects, but differing somewhat either in their phenomena or in the environment. The great obstacle to the development of a science of medicine is the difficulty in ascertaining what cases are sufficiently similar to be comparable, which difficulty is in its turn largely due to insufficient and erroneous records of the phenomena observed. The defect in the records is largely due, first to ignorance on the part of the observers; second, to the want of proper means for precisely recording the phenomena; and, third, to the confused and faulty condition of our nomenclature and nosological classifications.[53]

Even though observation had become "objective" through the use of precision-measuring instruments, medical inference was still predicated on professional clinical judgment.

Like these physiologically oriented clinicians, the research physiologist shared this bifurcated view; "objective" data could be obtained about the physiological organism even though generalizations about that data still depended on the professional aptitude of the individual researcher. To a significant extent, this antipathy toward formal "objective" rules of inference derived from the laboratory orientation of the physiologist. Such laboratory training inculcated in the physiologist the view that he/she was uniquely qualified to interpret physiological data; any appeal to formal and quantitatively "objective" rules risked undermining the physiologist's professional legitimacy.

The issue was nicely highlighted by the German physiologists' rejection of the mathematical view of medical statistics promulgated by the mathematician and physicist Gustav Radicke (1810–1883), a professor of physics and meteorology at the University of Bonn. In an 1858 article in Wunderlich's journal, the *Archiv für physiologische Heilkunde*, Radicke attempted to point out what he saw as the mathematical ignorance of his medical contemporaries. His stated purpose was to help in "stemming the stream of baseless and, to a great extent, erroneous doctrines which daily threaten to overwhelm medical science."[54]

The medical context for Radicke's analysis was the increasing number of attempts in this period to use quantitative techniques to establish the chemical constituents and volume of the urine. These techniques drew their inspiration from Justus von Liebig and his followers, who believed that urine analysis was a sensitive and significant indicator of generalized vital chemical operations, including the effects of therapeutic procedures. These issues were brought to Radicke's attention by his fellow Bonn physician Friedrich Wilhelm Boecker.[55]

Although Radicke did not refer to Louis directly, he began his analysis by observing how some physiological researchers used something akin to the numerical method. An individual patient would be given a particular therapy for a period of days, with all other aspects of that patient's life held as constant as possible. The daily excreta, most notably the urine, would then be analyzed from a chemical standpoint. The data obtained would be compared with the results from a second series in which either the same patient or some other patient would not be given the therapy in question. By a direct comparison of the arithmetic means for the various values obtained for the first series as compared with the second, the researcher decided whether or not the therapy had been effective. Radicke maintained that the principal problem with this method of direct comparison was that "those who employ it [the method of direct comparison] do not clearly comprehend the real significance and value of Arithmetic Means, and, as a consequence, have attributed to them a value which in such applications as these they do not actually possess."[56]

Radicke gave examples of several kinds of mean values that could be computed, such as an arithmetic mean, a geometric mean, a harmonic mean, and finally what Radicke dubbed a quadratic mean, which was defined as the square root of the arithmetic mean of the squares of the given numbers in the series.[57] He illustrated how the concept of a mean could be used in medical research by analyzing two series of researches published in the medical literature on the effects of sarsaparilla on urination. A patient's diet was kept constant except that, in the first series of twelve days, the patient was given a certain quantity of sarsaparilla daily, while, in the second series of twelve days, distilled water was administered instead of the decoction of sarsaparilla. In the first series (with sarsaparilla), the quantities of urine passed daily in cubic centimeters were 1,467, 1,744, 1,665, 1,220, 1,161, 1,369, 1,675, 2,129, 887, 1,643, 934, 2,093. In the second series (without the sarsaparilla), the quantities of urine passed daily in cubic centimeters were 1,263, 1,740, 1,538, 1,526, 1,387, 1,422, 1,754, 1,320, 1,809, 2,139, 1,574, 1,114.

By taking a simple average, one might assume that sarsaparilla had some effect on reducing the excretion of urine, since the average for the first series was approximately 1,498.9 as opposed to the average of 1,548.9 for the second series. Radicke cautioned against such hasty generalizations, noting that if the experiment had been stopped on the eighth day in each case, the average for the first series would be 1,554 and the average for the second series would be 1,494, which would imply a contrary result. Radicke attributed these erratic results to such variable factors as atmospheric agencies (temperature, moisture, pressure of the air), bodily or mental alterations, individual idiosyncrasies, or rapid fluctuations in the weight of the body.[58]

Radicke contrasted the accuracy of observation in the science in which he had been trained, astronomy, with the observations of the physiologist. He noted that recently developed instruments for angular measurements had attained "so high a degree of perfection" that in cases such as geodetic operations "we may obtain an average whose accuracy shall sensibly surpass the limits of that of any individual measurement, or even of that of its smallest graduation."[59] In contrast to this situation in the passive or observational sciences such as astronomy or geodesy, Radicke pointed out the complexity of making a comparison in quantitative chemical analysis.

Radicke noted that every observation was tainted by a certain error; the extent of this error was determined according to the "rule of probabilities." In forming his estimate of the error, Radicke advocated determining the mean of each observational error. The most appropriate mean value to be utilized in this analysis was the quadratic mean (the arithmetic mean would be of no practical value, since the sum of the differences of each observation from the mean value would produce a null outcome). Radicke did not go into a detailed account of the reasons for selecting the quadratic mean as the mean of choice in reducing observational error, because "it would presuppose . . . an amount of mathematical knowledge which, in the circle of readers for whom this treatise is designed, is not to be expected."[60] Nevertheless, he did note that because squared quantities are always positive, both positive and negative errors would be of equal value, and that the quadratic mean would demonstrate more unequal errors with a greater value.

After providing an overview of how to evaluate a series of numbers to minimize the possible errors in observation, Radicke observed that most series are insufficiently large to draw definitive conclusions. He provided suggestions for reducing error when dealing with such incomplete series. For instance, one should determine if an excessive disturbance has come into play which could produce an illusory result.[61]

Radicke then examined the specific case of the supposed influence of sea air and sea water on the metamorphosis of tissue. He appealed to the results of the professor of pathological anatomy and general physiology at the University of Marburg Friedrich Wilhelm Beneke (1824–1882), who had argued that sea-bathing simultaneously increased the excretion of urea and sulfuric acid and diminished the excretion of phosphoric acid. Beneke justified his conclusions by comparing the averages for the amount of urine, urea, sulfuric acid, and phosphoric acid excreted from patients at a seaside resort (Wangeroge) with the amounts from patients who were not at a seaside resort (Oldenburg). He found the increase in the amount of cubic centimeters of urine to be 152, the increase in the urea to be 3.1 grams, the increase in the sulfuric acid to be .28 grams, and the decrease of the phosphoric acid to be .51 grams.[62]

Radicke criticized these results, observing that the means obtained were in some instances derived from only four observations. He continued:

> The most cursory glance will show how little inference could be drawn from these very unequal numbers as to what the following ones would have been if the observations had been continued some days longer, and therefore, what small ground there is for assuming that any succeeding numbers might not have entirely inverted these relations.[63]

Radicke observed that the particular type of mathematical reasoning he had proposed was not held in high regard by the contemporary medical community. He hoped that his analysis had

> exposed, in a way that will be intelligible to medical readers who are not proficient in mathematics, the principles upon which conclusions should alone be deduced from any given series of observations, and that I have placed them in a position to deal with any observations they may make, in such a manner as to obtain certain results. Unfortunately, in the numerous pharmacological and balneological papers which, in ad-

dition to those cited above, I examined, hardly a tithe of the results published can be looked upon as certain. It will, therefore, give me much satisfaction if this contribution of mine should attain its object, and should be of some use in developing the important branch of medical science, of which investigations of this description are the basis.[64]

Radicke's intended audience, physiological researchers, were generally not receptive to the methodological concerns that he had addressed. Like the clinical critics of Louis and Gavarret, the research physiologists believed that their formal medical training ensured that they were uniquely qualified to interpret their own data. Just as the clinicians and epidemiologists responded to Gavarret, so likewise did the physiologists respond to Radicke: his approach was an unorthodox procedure that could be dismissed as unnecessary by the trained physiologist.

One of the first medical researchers to reject Radicke's professional competence to judge physiological data was Karl Vierordt of the University of Tübingen. Vierordt was no stranger to quantitative reasoning. In 1852 he had developed a technique of counting red blood cells by diluting a measured quantity of blood and assessing the number of cells in a volume of that dilution with the aid of a microscope. Vierordt reported his findings in Wunderlich's *Archiv für physiologische Heilkunde*; he even applied the "numerical method" of taking averages (although he did not attribute the technique to Louis)—after counting the blood cells in his own blood on nine different occasions, he found that his average blood count was 5,714,400 corpuscles per cubic millimeter. Although Vierordt's experimental method was too arduous to be applied on a regular basis, it demonstrated his willingness to apply quantitative reasoning in certain biomedical contexts.[65]

Nevertheless, Vierordt distrusted the application of formalized methods of inference to medical data, favoring instead what he called the logic of facts. Vierordt framed his discussion in terms of a survey of the various methods of employing statistical reasoning. The one most commonly used involved simply collecting numerical results, giving greater weight to a result if a larger collection of statistical instances was produced (in other words, the method of Louis even though Louis was by then no longer mentioned by name). The second method analyzed the numbers generated from the standpoint of the calculus of probabilities (the method of Gavarret, who was mentioned in passing). Vierordt characterized such probabilistically based conclusions as "purely formal"; that is, they depended entirely on mathematical formalism and did not recognize extenuating circumstances unique to the individual case. The third method recognized the value of such mathematical treatment but also recognized that "there may be other and independent reasons, arising from the nature of the object itself, which may, and even must be, regarded in drawing a conclusion."[66]

Vierordt did not deny the importance of the calculus of probabilities; rather, he objected to the exclusive reliance on such mathematical formalisms when the "logic of facts" dictated otherwise. The informed medical researcher should rely on both mathematical reasoning and empirical evidence. Vierordt maintained that in certain instances "to supplement the formal deficiency of the researches, grounds of conviction from the nature of the case itself come into operation which are not amenable to the logic of the calculus of probabilities."[67] For Vierordt, the calculus of probabilities was mathematically correct, but not always applicable to the particular medical concerns at hand.

Radicke's article generated a similar response from F. W. Beneke, the individual whose published data Radicke had analyzed. Like Vierordt, Beneke framed the issue largely in terms of a disciplinary dispute. He began by asking what the competency of mathematicians and physicists was that permitted them to be "authoritative critics of medical and physiologico-chemical matters, unless they are themselves to a certain extent physiologists." Like his fellow physiologist Claude Bernard, Beneke lauded the general attempt to reduce the phenomenon of vital processes to a kind of numerical or mathematical order; however, he maintained that "sound physiological discrimination" was superior to purely mathematical investigations.[68]

Since the human organism was not like an inanimate machine, such as a steam engine, and was constantly subject to myriad influences on the various vital processes, it was necessary to engage in physiological experimentation rather than formal mathematical analysis of various numerical averages that had been observed and measured. Beneke referred to a kind of experimental intuition acquired by a researcher over time which was as reliable as the results obtained from numerical analysis:

> Every one ... who has carried out many investigations of this kind, knows that during the progress of his experiments a confidence or mistrust in his results arises, which is much more convincing than the numbers themselves, and the experimenter is often without doubt as to the correctness of his result, notwithstanding the demonstration of the mathematician that the difference which may have been found between the two Means obtained from the series under examination, lies within

the limits of error of the calculation. . . . The physiologists, as well as the medical man, will look upon all the numerical values connected with these investigations as only approximately correct, and . . . the most careful estimate of probabilities will not rectify the numbers, if proper physiological considerations have been deficient, or if careful manipulation and processes were not adopted in obtaining them. . . . However accurately he [the mathematician] may determine the Mean of a given series of numbers, he works with quantities whose value he himself cannot estimate; and he may in the end evolve a result as mathematically correct, which is probably altogether false, in consequence of its being founded upon individual observations which are faulty.[69]

The role of the mathematician was of secondary importance in physiological researches to the experienced observation of the physiologist or the physician, "who know how to estimate how far the way in which the investigation is conducted, the numbers obtained, the conclusions drawn, and, still more, the ruling ideas and general cast of mind of the experimenter, are deserving of confidence, and correspond to our previous knowledge."[70]

Radicke responded to his physiologist critics. On the largely professional concern that he lacked the requisite physiological knowledge to judge the issue, Radicke maintained that his conclusions would follow regardless of what objects were being compared:

The fact is, that all that it is necessary to know, in laying down a rule for the comparison of two series of observations is—what we may learn from the series themselves, viz., whether fluctuations exist in those series, and if so, how great those fluctuations are. An acquaintance with their causes is quite unimportant.[71]

On the issue that there existed a "logic of facts" that could overrule the logical structures of mathematical theories, Radicke countered that the rules he had laid down "in their origin . . . are not purely mathematical, but that they contain a certain arbitrary element."[72] He referred to the "slight fluctuation" in the results about a mean value and observed that a researcher could never obtain absolute certainty but only a degree of probability. In an observation that would prove to be particularly prescient in light of later German commentaries on Gavarret, Radicke declared, "One person may place greater reliance in the certainty of the result than another would, which depends upon the undefinable nature of the expression, 'sufficient probability.' "[73] For the difference between two mean values not to be attributable to chance, Radicke had assumed that the mean difference should at least exceed the mean fluctuations. Although such an assumption would require a rather high degree of certainty before conclusions could be accepted, Radicke still believed it was necessary because

> *such large superstructures are erected* on physiological and pharmacological conclusions; and because new theories, or perhaps new methods of treatment, may be founded upon these conclusions, whose effect, both upon science and practice, is restrictive in proportion as those theories are considered to be well founded, whilst it is too easily forgotten that they are based upon data that are only more or less *probable*, rather than *certain*.[74]

The impact of Radicke's attempt to formalize the processes of generalization for the German physiologist paralleled Louis's and Gavarret's earlier attempt to formalize the processes of generalization for the French clinician. Both were largely programmatic statements that were subsequently debated but never put into practice. In part this derived from the nature of both works, which were isolated accounts; Gavarret and Radicke pursued their respective careers in physiology and physics after producing their analyses of medical statistics. More fundamentally, however, the rejection of Gavarret and Radicke (by clinicians and physiologists, respectively) reflected a continuity in medical outlook even though the focus for understanding disease had shifted from pathological anatomy to physiology. For the medical researcher, professional expertise continued to be viewed as the basis of the inductive generalization process, regardless of whether that expertise was carried out in the clinic or at the laboratory bench; it was predicated on the "tacit knowledge" of the individual researcher.[75]

As in the case of Gavarret, the views of Radicke were never systematically put into practice in the course of the nineteenth century. The anonymous reviewer of Hirschberg's work in the *British Medical Journal* of 1875 believed that such an account of medical statistics would be useful because "it is now some years since these latter [the views of Radicke and his critics] were published."[76] Radicke had rapidly become a distant memory.

What was significant about these debates, however, was the manner in which the dichotomy between informed professional judgment and statistical objectivity, as posed initially by the debates surrounding Louis's numerical method, had survived even though the medical context had shifted from the perspective of the clinical physician to that of the physiological researcher (and the geographical context had shifted from France to Germany). Louis had favored statistical objectivity at the level of the population because he desired that medicine acquire "scientific" standing. His critics had empha-sized the professional judgment of the clinician because they saw medicine as primarily an "art." Claude Bernard then effectively turned Louis on his head: he accepted Louis's vision of medicine as a science but saw the science of medicine as focused on the individ-ual, physiologically deterministic organism. Bernard's analysis was intimately related to his vision of making physiology a scientific disci-pline autonomous from clinical practice. Some prominent clinicians such as Wunderlich tried to steer a middle course between Louis and Bernard: they remained in the clinic as Louis did but attempted to understand disease in physiological terms the way that Bernard did. For these individuals, the patient had become a mass of quantifiable physiological data, and generalizations over a population, as prac-ticed by Louis, had become data analysis. The patient had been ob-jectified, but the data still required interpretation by the profession-ally trained individual researcher. However, as the negative response to Radicke indicated, any attempt to apply formalizable categories of inference to experimentally determined data (either in the clinic or the laboratory) fell largely on deaf ears; the "logic of facts" could still overrule the prestige of quantification. Such a situation was going to change, albeit very slowly, with the birth of the biometrical school in late-nineteenth-century Great Britain under the auspices of Francis Galton and Karl Pearson. They would provide both the cognitive and social structure that enabled the mathematically informed medical statistician to emerge as a new kind of scientific professional.

THE BRITISH BIOMETRICAL SCHOOL AND BACTERIOLOGY: THE CREATION OF MAJOR GREENWOOD AS A MEDICAL STATISTICIAN

THE LAST THIRD of the nineteenth century witnessed innovations that would fundamentally alter the manner of debating the role of statistical methods in medicine. These innovations were tied to an increasingly scientific self-image within both the medical and the statistical professions. Statistics was transformed from an empirical social science, concerned with the collection and description of social phenomena, into a mathematical applied science; it became a body of formal analytical techniques for analyzing aggregative data. This transformation derived from the British biometrical tradition associated with such figures as Francis Galton and Karl Pearson. In the medical profession, likewise, there emerged a group of individuals who argued for the use of "scientific" methods in both diagnosis and therapy. In the realm of diagnosis, precision-measuring instruments were seen as making observation objective and "scientific." In the realm of therapy, the laboratory-based methods of bacteriology were seen as a key to providing medical treatment.

These multiple "scientific" approaches to both statistics and medicine intersected during the first decade of the twentieth century in a series of debates in Great Britain. Unlike earlier discussions of statistical methodology, however, these intellectual exchanges were carried on by two self-sustaining communities, each of which viewed itself as a scientific subdiscipline. The conditions were ripe for the emergence of a new type of professional medical statistician who understood both medicine and mathematical statistics.

In order to provide a broader context for these social and intellectual transformations, it is necessary to discuss the formation of new scientific professions. Recent commentators have noted that in order for new professional specialties to emerge within science, in addition to new ideas or techniques, several events are necessary: a network of scientists interested in the new field must be formed; means of communication between them, both formal and informal, must be established; a mechanism must be devised for recruitment to, and training in, the field, which must be given some stable form; and sufficient financial and other resources are needed to maintain the new field.[1] Several of these patterns of social reorganization were occurring within both the statistical and medical communities in Great Britain at this time.

The major theoretical innovations that have often been seen as the foundation of modern statistical theory derived from the intellectual insights of Francis Galton (1822–1911). Galton studied medicine at Cambridge but achieved no great distinction. Upon receiving his inheritance, he explored Africa during the period 1850–52 and received the gold medal from the Royal Geographical Society in 1853 in recognition of his achievements. In the early 1860s he turned his attention to meteorology, publishing one of the first attempts to explain weather phenomena using graphical methods in his 1863 work *Meteorographica*. Although Galton subsequently published on such varied subjects as psychology, anthropology, sociology, education, and fingerprints, the dominant concern in his work after 1865 was the study of heredity.[2] In Charles Darwin's 1859 work *On the Origin of Species*, Galton found the inspiration both for a new vision of the role of science in society and a key for major innovations in statistical theory.

Galton saw in evolution a way to reconstruct society in the image of science. He called for "the establishment of a sort of scientific priesthood throughout the kingdom, whose high duties would have reference to the health and well-being of the nation in its broadest sense, and whose emoluments and social position would be made commensurate with the importance and variety of their functions."[3] Such a priesthood was seen as necessary because of the increasing complexity of modern society.

This general faith in the methods of science can best be understood in the context of a late Victorian intellectual movement known as scientific naturalism. In addition to Galton, the movement numbered among its members T. H. Huxley, John Tyndall, Herbert Spencer, and W. K. Clifford. It had various tenets such as the view that the proper direction of human life itself derives from following the methods as well as the content of a new secular, scientific worldview. Scientific methods must receive wide dissemination among the educated public and the scientifically trained must become leaders of British intellectual culture.[4]

Galton's idea of a scientific elite governing intellectual culture was further reflected in his attitude toward statistics; he placed equal emphasis on its theoretical dimensions and on its widespread applicability. At the annual meeting of the BAAS for 1877, he advised the Council that Section F ("Economic Science and Statistics") be discontinued. He observed that during the years 1873–75 not a single memoir treated the mathematical theory of statistics, and that if any paper were submitted dealing with such subjects, "the proper place for it would be in Section A ['Mathematical and Physical Sciences']." Galton admitted that the statistical section dealt with important matters of human knowledge; however, he observed:

> Usage has drawn a strong distinction between knowledge in its generality and science, confining the latter in its strictest sense to precise measurements and definite laws, which lead by such exact processes of reasoning to their results, that all minds are obliged to accept the latter as true. It is not to be expected that these stringent conditions should be rigorously observed in every memoir submitted to a scientific meeting, but they must not be too largely violated; and we have to consider whether the subjects actually discussed in Section F do not depart so widely from the scientific ideal as to make them unsuitable for the British Association.[5]

The exclusion of section F from the BAAS was seen as an issue of professional standards:

> The Section is isolated and avowedly attracts much more than its share of persons of both sexes who have had no scientific training, its discussions are apt to become even less scientific than they would otherwise have been. On the other hand, any public discredit which may be the result of its unscientific proceedings has to be borne by the whole Association.[6]

William Farr provided a report to the committee of the BAAS from the secretaries of the Statistical Society. They acknowledged that more people with little or no scientific training attended the meeting of this section than others. Nevertheless, they maintained:

> In exchange for this evil [unscientific discussion], it must always be remembered, scientific men have a better opportunity in this section than in any other of communicating some notion of scientific method and its value, and of the conclusions of scientific study, to the unscientific multitude. If the British Association is to exist for the "advancement of science," it cannot but fulfil its end in making politicians and philanthropists generally aware of the necessity of scientific method and knowledge in their favourite subjects.[7]

Galton's interest in the intellectually elite members of society related not only to a professionalizing strategy to ensure the high quality of science but also influenced the formulation of his idea of statistical correlation. As several commentators have observed,[8] Galton was led to the method of correlation by his interest in the science that he termed *eugenics*, the evolutionary doctrine that the condition of the human species could most effectively be improved through a scientifically directed process of controlled breeding. He became interested in explaining what he believed was a curious, thoroughly regular and law-like phenomenon: the distribution of hereditary genius in gifted families. Drawing on Herschel's exposition of Quetelet's work, Galton applied the Gaussian law of error to human beings in his 1869 work *Hereditary Genius* and decided that intelligence must follow such a probability distribution. However, unlike Quetelet, Galton was more interested in distributions and deviations from the mean than in the average value itself. As a result, he observed that reversion toward mediocrity in successive generations was a mathematical consequence of the normal curve. That is, if a population were normally distributed, it could be deduced that in a second generation there would be a normal distribution about the same mean and dispersion, but one in which the exceptional members would typically not be descended from exceptional members of the previous generation. This conclusion constituted a decisive step in what Ian Hacking has recently called the taming of chance: Galton had regarded the normal distribution of traits as an autonomous statistical law. Where Quetelet had made biological averages into something "real," Galton had now added another tier to that reality: he had made correlations as real as causes.[9]

Galton's social vision of the scientist as a cultural leader and his intellectual innovations in statistical methods found a devout disciple in the person of Karl Pearson (1857–1936). As Stephen Stigler has observed, Pearson "lacked Galton's originality . . . but it was his zeal . . . that created the [statistical] methodology and sold it to the world."[10] Pearson was educated privately at University College School and at King's College, Cambridge, where he was Third Wrangler in the Mathematics Tripos in 1879. On leaving Cambridge, he spent part of 1879 and 1880 studying medieval and sixteenth-century German literature at Berlin and Heidelberg Universities. He then read law at Lincoln's Inn and was called to the bar by Inner Temple in 1881. During this period, he gave lectures on mathematics, philosophy, and German literature at the London Society for the Extension of University Teaching, the South Place Institute, and other societies and clubs devoted to adult education. He became professor of mathematics at King's College London in 1881 and at University College London in 1883. By this time Karl Pearson was lecturing on free thought, socialism, the woman's question, Karl Marx, and Ferdinand Lassalle, among other matters. In June 1884 he was appointed to the chair of Applied Mathematics and Mechanics at University College London and in 1891 he was appointed to the professorship of Geometry at Gresham College. This latter position entailed giving a series of twelve lectures a year and it was in this context that Pearson first demonstrated his interest in statistics, giving lectures with titles like "The geometry of statistics," "The Laws of Chance," and "Probability." At this time Pearson met the zoologist W.F.R. Weldon, also of University College, who was an early advocate of the use of Galtonian statistical methods in solving evolutionary problems. Through Weldon Pearson was introduced to Francis Galton, the individual who first freed him "from the prejudice that sound mathematics could only be applied to natural phenomena under the category of causation."[11]

Pearson demonstrated his loyalty to Galton's ideas both in his published works and in his institutional position as a professor at University College London. In print he maintained that empirically determined "facts" obtained by the methods of science were the sole arbiters of true belief. As he nicely summarized the issue in his popular work *The Grammar of Science*:

1. The scope of science is to ascertain truth in every possible branch of knowledge. There is no sphere of inquiry which lies outside the legitimate field of science. To draw a distinction between the scientific and philosophical fields is obscurantism.

2. The scientific method is marked by the following features:—(a) Careful and accurate classification of facts and observation of their correlation and sequence; (b) the discovery of scientific laws by aid of the creative imagination; (c) self-criticism and the final touchstone of equal validity for all normally constituted minds.

3. The claims of science to our support depend on: (a) The efficient mental training it provides for the citizen; (b) the light it brings to bear on many important social problems; (c) the increased comfort it adds to practical life; (d) the permanent gratification it yields to the aesthetic judgment.[12]

Pearson also argued in favor of the almost universal application of statistical methods:

The very statement of the law of causation involves antecedents—sameness of causes—which are purely conceptual and never actual. Permanence and absence of individuality in the bricks of the physical universe are only demonstrated in the same way that the bricks of a building are for many statistical purposes without individuality. The exact repetition of any antecedents is never possible, and all we can do is to classify things into like within a certain degree of observation, and record whether what we note as following from them are like within another degree of observation. Whenever we do this in physics, in zoology, in botany, in sociology, in medicine, or in any other branch of science, we really form a contingency table, and the causation of the physicist solely results from the fact—not that the contingency coefficient of everything physical is unity—but that he has so far worked to most profit in the field, where his contingency is so near unity that he could conceptualise his relationships as mathematical functions.[13]

Pearson more forcibly spread the view that statistical methods were a tool of scientific inference by drawing on the facilities provided by his university position. In 1892 he and Weldon began the collaboration that was to grow into the biometric school. Two years later Pearson offered his first advanced course in statistical theory, thereby inaugurating University College as the sole source for instruction in modern statistical methods before the 1920s.

To extend further the view that quantitative and statistical methods were applicable to the problems of the science of life, Galton, Pearson, and Weldon founded the journal *Biometrika* in 1901. This was a self-conscious effort to convince a largely indifferent scientific community of the value of mathematical statistics. Pearson wrote to Galton on the plan to found such a journal because the sort of statistical memoirs that he had in mind would "not find a place" in better-established publication outlets such as the Royal Statistical Society.[14] He later requested that Galton join the editorial board so that the journal could be "got off the ground" because "You [Galton] hardly know how much perhaps, of weight your sympathy expressed in some form carries with it, as perceived in America, and it will be an uphill battle for some time with the biologists."[15] Throughout these early planning stages, Pearson envisaged the journal *Biometrika* as a way to legitimate statistical methods as a standardizable category of scientific inference. As he wrote to Galton, "Half the Editors' work will be to show authors *gently* how to use their own data!"[16]

Galton responded to Pearson with both advice and financial support. He acknowledged that there would be hostility to the biometrical approach but maintained that "few new & good things are ever accomplished except in the teeth of violent opposition."[17] Also, Galton provided £200 to cover the initial cost of publication and donated the financial reward he received from the Darwin fund to help finance the journal.[18] In addition Galton offered a sizable endowment in 1904 to University College to promote "the *exact* study of what may be called *National Eugenics*" by means of "the higher methods of statistics." The endowment consisted of an initial outlay of £1,500 for three years to establish a research fellowship in eugenics for a period of three years with the possibility of making the endowment permanent at £500 a year.[19] By drawing on these sources of financial and institutional support, Pearson transformed his relatively informal group of followers into an established research institute, consisting of the Eugenics Laboratory and a "Biometric Laboratory."[20]

There were several sources of tension in the type of research institution that Pearson created with Galton's financial backing. For instance, Pearson claimed that the laboratory was justified because of its potential benefits to the nation (i.e., its practical value) but still tried to maintain that the research institute was above the political fray (i.e., committed to "pure" research). As he observed in a letter to Galton in 1909:

> We have got to convince not only London University, but the other universities that Eugenics is a science and that our research work is of the highest type and as reliable & sober as any piece of physiological or chemical work . . . that we are running no hobby and have no end in view but the truth. If these things can be carried out we shall have founded a science to which statesmen & social reformers can appeal for marshallable facts.[21]

Pearson was, in particular, attempting to distance his work from organizations like the Eugenics Education Society, which had been founded in 1907 to generate public interest in eugenics. As a founding member wrote to Galton, "Its purpose . . . is, on the whole, frankly propagandist."[22] In contrast to such overtly social reform movements, Pearson maintained, "All I want is to stand apart doing the scientific work, but not in any way hostile to the Eugenics Education Society, giving it any facts we can or an occasional lecture; but not being specially linked to it in any manner."[23]

One method that Pearson used to differentiate his "scientific" research from the "social" concerns of the Eugenics Education Society was to appeal to the use of mathematical statistical methods in his biometrical laboratory: "My own view is that our work is in different fields & is supplementary, but I fear the Eugenics Education Society will not accept this view, and does not fully grasp that we can be quite sympathetic, but must do our own work in the narrower field of statistical research."[24] Statistics became a kind of quantitative legitimating discourse by which Pearson could argue for the scientific underpinnings for his research into heredity.

Such a professionalizing strategy merely brought to the fore the problematic position of statistical arguments within contemporary scientific and medical reasoning. Even though Pearson's eugenic researches were based on the theory of evolution by natural selection,

Darwin himself had shown an antipathy toward the use of statistical arguments; his work was not available as a *locus classicus* to justify a statistical methodology. Pearson wrote to Galton asking if he had preserved any letter from Darwin that contained "any apt remark as to the need of statistical method in solving evolutionary problems?"[25] Galton doubted that Darwin "ever thought very much or depended much on statistical inquiry in his own work . . . though . . . he quotes statistical results that others had arrived at, not infrequently."[26] On consultation with Darwin's sons Leonard and Frank, Galton advised Pearson: "I fear you must take it as a fact that Darwin had no liking for statistics. They [Leonard and Frank] even thought he had a 'non-statistical' mind, rather than a statistical one."[27]

Pearson was thus placed in a rather tenuous position. Even though he relied on the interest in eugenics in late Victorian and Edwardian society (as reflected in the Eugenics Education Society) to justify his research, he used his statistical methodology both to assert his scientific credentials and to distance himself from the more overtly "propagandistic" statements made on behalf of eugenics. However, statistical methods themselves were seen as dubious by those members of the British scientific establishment Pearson was most interested in convincing, namely biologists and mathematicians. Statistical methods were still associated more with "social" analysis than with "pure" scientific research.

Throughout the first decade of his research into statistical theory in the late 1890s and early 1900s, Pearson often commented on the difficulty of finding a wider audience for mathematical statistical methods. As he wrote to Galton in 1898: "I fear . . . you are the only part of the scientific public, which takes the least interest in my work. The mathematicians look askance at any one who goes off the regular track, and the biologists think I have no business meddling with such things."[28] The founding of the journal *Biometrika* did not fundamentally alter such an indifference toward statistics among the British biological community.[29] In 1903 Pearson wrote to Galton:

> In Cambridge we have *two* subscribers, one a personal friend of my own & one of Weldon's, and people like Bateson when they want to criticize us borrow one of these copies! The battle is idle against such an attitude. . . . Neither mathematicians nor biologists will give an appointment to a biometrician & so one cannot encourage students who have to work for their living to do this sort of work.[30]

In 1906 Pearson even went so far as to threaten to resign from the Royal Society when it refused to publish Raymond Pearl's study based on a statistical argument that protozoa do not mate at random:

"My chief work & interest lies in biometry & the application of exact methods to biology. If the Royal Society will have nothing to do with it, and publishes whole series of papers and reports whose writers lack the most elementary knowledge of statistics, it ceases to appeal to me in any way."[31] When Galton counseled that he had a better chance of educating biologists in statistical methods by remaining in the society, Pearson planned to write the Royal Society that the Zoological Committee had rejected Pearl's paper "principally on the ground that they do not see the biological significance of the constant calculated," i.e., a correlation coefficient.[32] Although Pearson played his self-appointed role of salesman for statistical methods with great zeal, he still had to contend with widespread indifference from large segments of the scientific community.

Pearson attempted to build intellectual bridges with other professional pursuits such as medicine in order to further his and Galton's vision of the widespread application of statistical methods. Galton expressly urged that Pearson court support from the medical community: "I trust you will see your way to make a considerable part of the contents of the Journal [*Biometrika*] intelligible to those scientific men who are not mathematicians. It ought to be attractive to medical men and such like, also to statisticians of the better kind."[33] Pearson took Galton's advice to heart. His attempt to distance his work from that of the Eugenics Education Society was motivated, in part, by the fact that he "could not get the help we are getting from the medical profession from pathologists or physiologists, if we were supposed to be especially linked up with these names. Rightly or wrongly it would kill Eugenics as an academic study."[34] The Eugenic Council was seen as a hindrance in the attempt to win the support of the medical profession:

> The medical men are coming in and giving us splendid material ... often confidential and personal histories. But Havelock Ellis, Saleeby & others on the Eugenic council are red rays to the medical truth, and if it was thought we were linked up with them, we should be left severely alone. I think it a very great thing to have won even partial confidence from a portion of the medical world, and if we can keep it & extend it, we shall have really done a great stroke in forwarding the scientific side of Eugenics.[35]

The receptivity of the medical profession to such biometrical statistical methods can best be understood in the context of a broader debate over the proper role of scientific methods both in the training and the daily practice of medicine. Among clinicians there were antithetical responses toward scientific reasoning. On the one hand, there was the London-based medical elite who invoked a rhetorical claim advocating "clinical science" but actually placed heavier emphasis on classical or general education than on a narrowly technical or scientific one and praised the attainment of character more than the pursuit of expertise. On the other hand, there were those clinicians who advocated the use of tools derived from the physiological sciences to investigate disease and generally espoused the introduction of continental scientific methods into the British medical curriculum.[36] In contrast to both of these clinical approaches were the laboratory-based methods of the bacteriologist.[37]

The attitude toward statistical reasoning of each of these segments of the medical profession could be seen as a function of its general view of science and its more specific view of the role of statistical reasoning within science. Clinicians who continued to emphasize the "art" of medicine conceived of statistics as their predecessors had in the time of Louis (direct comparison) and as adding little information beyond that supplied by experience. Those who argued that there existed a "clinical science," basing diagnosis on physiological instruments, saw statistics as a way to make observation more objective and "scientific." Finally, bacteriologists agreed with other laboratory researchers (e.g., Bernard)—that statistics might provide insight but did not constitute "scientific" evidence.

One representative physician who continued to emphasize the "art" of medicine was Philip Henry Pye-Smith, a fellow of the Royal College of Physicians and a consulting physician to Guy's Hospital. In an address to the annual meeting of the British Medical Association in 1900, he argued for the primacy of pathology in medical diagnosis and claimed that the practice of medicine was an art based on experience rather than an abstract science: "We must never allow theories, or even what appear to be logical deductions, or explanations however ingenious, or *statistics* [my emphasis] however apparently conclusive, or authority however venerable to take the place of the one touchstone of practical medicine, experience."[38] Even though Pye-Smith explicitly referred to "Louis and the French School," who had pioneered a statistical approach to medicine, he still maintained that there are "many pitfalls in deducing conclusions from statistics." For statistical inferences to be valid, the results had to be based on cases that were both numerous and accurate.[39]

Among the clinicians who advocated the ubiquity of scientific methods of explanation, one of the most prominent was the Regius Professor of Physic at Cambridge University, Clifford Allbutt (1836–1925). He had always wanted medicine to acquire a scientific character—he had chosen medicine as a career after reading Comte's *Cours*

de philosophie positive. During the 1870s he advocated the use of precision instruments in medicine and developed one of the first short
clinical thermometers as well as publishing an epoch-making monograph on the use of the ophthalmoscope.[40] By the time of his appointment to the professorship at Cambridge (1892), he could lament
the view that the clinician had remained the cultured generalist
rather than becoming the scientific specialist: "Academical or university methods—the theoretical habit of mind—will take an ever-
increasing place in education. It will more and more be the duty of
mankind to make itself theoretical. Why then the prevailing distrust
of theory, especially in England?"[41] In response to the charge that
such academic and intellectual pursuits might prove useless in the
realm of concrete practice, Allbutt cited the example of how "the vast
structure of Cambridge mathematics" as developed by Kelvin and
Stokes was "creating modern physics."[42] He lauded the pursuit of
university studies as important for the advancement of knowledge in
all areas of human understanding including medicine because

> without the incessant converse with theory no mind can be trained in
> the scientific habit—that is, in the cunning of nature. The man of scien-
> tific habit apprehends quickly the fashion after which nature at a certain
> juncture is likely to work; the apprentice who has not acquired this ca-
> pacity, however efficient a performer within the lines of his traditional
> handicraft, becomes a laggard even in practice, is always wasteful in his
> means, and is both inept and unwilling to readjust his methods in the
> light of advancing knowledge.[43]

As part of his espousal of a scientific approach to medicine, Allbutt
used statistical constructs in the diagnosis and nosological classification of disease. In an introduction that he wrote for an encyclopedia
entitled *A System of Medicine* (1901), Allbutt observed:

> By a typical case, then, we ought to signify (for we use the word very
> inconsistently) a case in which the symptoms do not differ largely from
> those occurring in the "morbid mean." As parallel instances, we may
> take the variation of the stature of men of a nation, about a mean, or the
> distribution of bullet marks on a target.
>
> Not only this question but many others also might be explained if by
> plotting out measurable symptoms in curves we could get the mean
> intensity of each and the amount of its variability; and could determine
> whether the measures are symmetrically arranged about this mean. To
> form such a curve the measurements would be set along the abscissa,
> and the numbers of instances as ordinates. This is, however, too difficult
> an undertaking to discuss here even were I capable of its discussion. We

have also to bear in mind that the treatment of statistics is somewhat dangerous, unless, carried out by one who has some acquaintance with the theory of errors; the curves might be constructed accurately, but they might be made and used on wrong principles.[44]

He observed that medical judgment is based "in the last analysis, upon statistics, and the result has the more validity as the number of observed cases increases."[45]

Allbutt's views indicate that, among those clinicians who desired to give medical diagnosis scientific credentials, statistical methods could find advocates. He represented the concepts of disease and health in terms of statistically constructed "norms" or standards.[46] Also, he portrayed statistics as an autonomous body of analytical techniques for handling aggregates of numerical data. He even viewed statistical results as capable of interpretation or further analysis without an intimate knowledge of the circumstances under which they were obtained—what Zeno Swijtink has called the "objectification of observation" in the nineteenth century.[47]

Despite programmatic statements by clinicians such as Allbutt on the usefulness of statistical methods, the principal focus for early-twentieth-century debates on the relationship between statistics and medical certainty derived from the rise of bacteriology, for a variety of reasons. Bacteriology offered the prestige of laboratory science in the diagnosis and treatment of disease. Statistics, in turn, offered the prestige of number and quantification in proving the efficacy of results. Furthermore, the introduction of new therapies developed in the laboratory implicitly raised the problem of statistical inference, that is, the extent to which the results were a product of chance given that the experimental therapies were tested on only a segment of the total population. Nevertheless, both the bacteriologist and the statistician still had to contend with the outlook of the clinician, who maintained that the bedside should remain the proper place for the diagnosis and treatment of disease. Such a "triangular" demarcation between the clinician, bacteriologist, and statistician over who should remain the final arbiter of medical knowledge formed the broader context for a series of debates between the bacteriologist Sir Almroth Wright (1861–1947) and the biometrical school in the first decade of the twentieth century.

Wright's early career nicely highlighted the growing prestige of bacteriology and the methods of the laboratory. After receiving his initial education at the Belfast Academical Institution and Trinity College, Dublin, Wright was awarded a traveling scholarship of £100 in 1883 to spend a year studying at the University of Leipzig, where he worked under such pioneers as Julius Cohnheim, Carl Weigert, and Carl Ludwig. Subsequently, Wright worked in Michael Foster's physiological laboratory at Cambridge, spent additional time studying in Germany, and eventually took the position of professor of pathology at the Royal Army Medical College in Netley in 1892.[48]

While at Netley, Wright developed a prophylactic immunization serum against typhoid fever, testing it on himself, his friends, and students attending his classes at Netley. Further tests were carried out in 1897 among nurses and attendants of the hospital for mental illness at Barming Heath, on the occasion of a severe typhoid epidemic. Wright then began to introduce his method of immunization into the army through the assistance of the Royal Army Medical Corps by immunizing soldiers in transit to India; he continued to immunize soldiers until the end of the Boer War in South Africa. An advisory board was then appointed to assist the War Office on questions of medical science. In 1902 the board recommended that the practice of antityphoid inoculation be suspended largely because of personal animosity between Wright and the board members. When Wright protested, a series of committees were appointed to decide on the merits of his statistical results.[49]

In 1904 Wright published *A Short Treatise on Anti-Typhoid Inoculation*, in which he spelled out both the concrete results of his research and the theoretical assumptions underlying his conclusions. He noted two general requirements "which concern the cogency of statistical evidence," namely that there exist a control group "which ought to correspond with the inoculated group in all points save only in the circumstance of inoculation," and that exact numerical records should be kept for both groups of the percentage incidence and case-mortality.[50] Despite such an appeal to statistical methodology, Wright warned that it was generally more advisable for the physician to accept the general conclusions implied by statistics than to worry excessively about the theoretical underpinnings:

> Persons who desire to wait till the evidence is obtained which satisfies the theoretical canons set forth at the outset of this statistical discussion are persons who desire to wait indefinitely. The plain, everyday man will find it possible to reconcile the demands of his statistical conscience with the demands of practical life. He will neglect the mint and anise and cummin of statistical criticism while holding fast to weightier principles of the statistical law.[51]

Wright specified conditions to determine when to accept statistical conclusions even if the collection of the data had been flawed. In general, small errors that did not undermine the final conclusion could be overlooked as could results based on a large number of figures because "all chance errors ... are spontaneously eliminated" by the excessive number of observations. However, Wright warned that if an error operated always on behalf of one conclusion as opposed to another, the error would not be eliminated by the accumulation of figures; e.g., the introduction of pseudo-typhoid cases into a study would undermine the result.[52]

Since Wright based his discussion of the efficacy of antityphoid inoculation on statistics, Lt.-Colonel R.J.S. Simpson from the advisory board of the War Office sought the assistance of Karl Pearson to help establish whether or not Wright's results were "significant." Pearson advocated a kind of clinical trial recommending that every alternate man in a regiment of eight hundred be inoculated so that "we can exhibit your results in correlative form, showing a distinct relation between inoculation and immunity."[53] Although Simpson maintained that the voluntary nature of the procedure precluded the indiscriminate inoculation of every second man, he did request that Pearson present his results to the committee and acknowledged his "expert" assistance: "Your opinions ... are extremely valuable, not only as those of an expert, but as those of an unbiased critic, while those of us who have been working at the subject are more or less prejudiced one way or the other."[54]

Pearson published the results of his study in the *British Medical Journal* in November of 1904. He determined mathematically the correlation coefficient between immunity and inoculation and between mortality and inoculation. According to these methods, pioneered initially by Galton, the coefficient of correlation (usually denoted by the letter r) provided a measure of the degree of relationship between two distinct classes of phenomena. If the phenomena were totally unrelated, then $r = 0$; if the phenomena were totally correlated, then $r = 1$. For most conclusions based on statistical comparison (such as the relationship between inoculation and immunity), the correlation coefficient would vary between these extreme limits.[55] Pearson hoped to use the correlation coefficient to see how antityphoid inoculation compared with other types of therapeutic procedures. For instance, he noted that the "protective character of vaccination as against mortality after incurring small-pox is very substantial" with a correlation coefficient of approximately 0.6. In addition Pearson noted that the correlation between recovery and administration of antitoxin in diphtheria was 0.47, and between the need for a tracheotomy and the administration of antitoxin, 0.24. For typhoid fever, in contrast, the average correlation between immunity and inoculation was found to be about 0.23, with individual results ranging from as high as 0.445 to 0.021. Pearson concluded that the results were significant; however, he was struck by the "extreme irregularity and the lowness of the values reached."[56] He speculated that environmental factors might also contribute to susceptibility to typhoid fever. In general Pearson declared that the results adduced to support Wright "certainly appear to fall into that range of intensity which would justify suspension of the operation as a routine method"[57] and advocated further study of the issue.

The editors of the *British Medical Journal* agreed with Pearson's criticisms of Wright. After giving a brief account of the series of committees that had studied the merits of Wright's statistics, the editors concurred that the effectiveness of immunization had not been conclusively established; therefore, it should not become a routine practice within the army. They regretted that a newspaper "of such influence" as the *Times* had come out in favor of Wright's position in an attempt "to force on the War Office and on the army a method which, on the testimony of its author, requires the labour of years to perfect, and the use of which involves possibilities of serious mischief."[58] The editors waxed literary with Homeric references declaring that "Dr. Wright may be pardoned for believing that none but himself can bend the bow of Odysseus, and he may be congratulated on having found a faithful Penelope in Printing House Square."[59]

Wright responded to the views of the editors and Pearson in terms of the issue of professional competence. As a physician trained in the laboratory methods of bacteriology, he did not wish to have his results criticized by a mathematician. Responding with more classical allusions, Wright dubbed Pearson a kind of "statistical Rhadamanthus before whom all the deeds of inoculators done in the body are destined to come up for judgement."[60] He conceded that Pearson's mathematical principles were both "unerring" and "beyond my intellectual ken." Nevertheless, Wright affirmed that "the standard of perfection which he [Pearson] exacts is immeasurably above the standard to which we can hope to attain in connexion with immunization against typhoid fever." In addition, he charged the advisory board with hiding "behind Professor Pearson's petticoats" in their attempt to keep antityphoid inoculation from being reintroduced.[61]

The debate continued with each protagonist framing the issue in terms of professional expertise. Pearson claimed that "any inoculation process must ultimately come before the calm tribunal of a statistical inquiry."[62] Wright maintained, in contrast, that "if Professor Karl Pearson had any practical acquaintance with medical matters he would not attach faith to any case-mortality based on a series of only 35 cases, and he would not see something 'most mysterious' in the false registration of a couple of deaths in a beleaguered city."[63]

Pearson used the debate with Wright as an occasion to plead that the medical profession become proficient in mathematical statistical methods. He stated that there was a "crying need for a more exact treatment of statistics in medical science" and advocated either the creation of a corps of expert medical statisticians or a "more friendly relation ... between the medical man and the professed statistician."[64] Pearson described the early attempts at such a union between medicine and statistics by the staff in his laboratory and the medical practitioners trained there. Wright reacted to the claim that medicine should be statistically grounded by invoking a maritime metaphor: "It is exactly as if a calculator, whose office it was to compute the strength of certain shipbuilding materials, were to contend that he, and not the practical seaman, was the proper judge of the performances of a ship."[65] Pearson responded in kind:

> The calculator would be a better man than the practical seaman to determine as a result of his calculations whether the loads the seaman insisted on the vessel carrying would not cause it to "hog" badly when launched. I opine that the ship "Routine Inoculation," if built on the present design, stands a fair chance of hogging badly if she be ever launched.[66]

Although Wright's professional vision of medical research as laboratory-based led him to downplay the statistical issues that Pearson advocated, his subsequent research into vaccine therapy would provide more grist for Pearson's statistical mill. Wright had begun this research when he was appointed to the position of pathologist and bacteriologist at St. Mary's Hospital, Paddington, in 1902. He believed that a dose of vaccine given during the course of an illness caused by a particular microbe might induce the formation of more antibodies and so help to *cure* the patient; vaccines composed of dead bacteria could be used to cure as well as to prevent disease. Wright and his associate, Captain Stewart Douglas, developed a quantitative measurement called an opsonic test to estimate the number of antibodies in the blood. They discovered that there was a substance in blood serum (opsonin) which prepared bacteria to be ingested by the leucocytes (white blood corpuscles). In a normal person the amount of opsonin remained constant, but in persons attacked by a bacterial substance, there soon resulted either an increase or a decrease of the opsonin in the blood serum. Usually, after the injection of a dose of a vaccine, there was a temporary decrease in the opsonic substance followed by a considerable increase. The opsonic power of the patient's serum was compared with that of a normal person to determine what was called the opsonic index.

Equal quantities of the patient's serum, a suspension of the microbes, and a suspension of normal leucocytes were mixed together and kept at body temperature for half an hour. Then a preparation of the mixture was made on a glass slide, stained with an appropriate dye, examined under a microscope, and the number of microbes seen in each leucocyte counted. From 25 to 100 leucocytes were examined and the average number of microbes in each leucocyte noted. A similar procedure was then carried out on a normal person as a control. The ratio of the first average count to the second constituted the opsonic index. Experiments determined that if the normal figure were taken as 1, then the normal variation seldom went below 0.8 or above 1.2. When the index was found to be above or below those limits, infection with that microbe was usually found. By these means, infections could be detected before the person concerned had shown any symptoms of the disease.[67]

On the basis of these researches, Wright hoped to revolutionize medical therapy. Vaccines composed of the dead bodies of the infecting organism could be used to stimulate the natural resistance of the body. As Wright proclaimed in a lecture of 1902, "The physician of the future will, I foresee, take upon himself the role of an immunizator."[68] Later Wright would contend that the role of bacteriological methods for medicine would soon require that "the physician . . . be a trained laboratory worker."[69]

Wright's support for the opsonic index and vaccine therapy can best be understood in the context of his broader vision of the function of "science" in medical research. Like Bernard in the nineteenth century, Wright maintained that the methods of the research laboratory were key to making advances in medical knowledge. As he observed in a 1905 newspaper article entitled "The World's Greatest Problem":

> It will, perhaps, now come home that after food and shelter and external defence and the administration of justice have been provided, the most urgent need in every civilized community is the need for medical research. It is for every far-seeing man to reflect upon the problem as to how the proper kind of workers can be enlisted for this work, and how the proper training and skilled direction can be provided for the workers. Until the economic conditions shall have been so altered as to attract to the work of research a proportion of the best ability in every country, the task of providing for that scientific skilled direction of medical research will be the really difficult problem.[70]

Wright heaped disdain on other approaches to combat disease focusing in particular on preventive medicine. He declared that "the sanitarian of to-day goes about his work with scales of ignorance over his eyes." For Wright, the claims that an outbreak of plague depended on filth, or was spread by rats, "rest on no positive knowledge." Even Ronald Ross's demonstration that malaria was spread by mosquitoes had not prevented the outbreak of the disease "by the most strenuous measures directed to the destruction of the external sources of infection." Wright concluded, "That dream of preventive medicine must not hold us back from seriously attacking the study of the processes of disease in the human organism."[71] Wright's research into vaccine therapy can largely be seen as an attempt to prove his broader claim that medical advances derived from laboratory research.

Those members of the medical profession who did not engage in research responded to Wright's arguments with reservations. An early commentator on Wright's proposals, clearly reflecting the perspective of the clinical physician, stated:

> We do not find his opinion of the position sound or his arguments free from fallacy. It is obvious, too obvious to be worth saying, that much of the recently acquired knowledge as to the production of disease by invading organisms is in its infancy. There open out from what we have learned many fields of inquiry which must be diligently traversed before the full value and the complete application even of what is now known will be in our possession. In the case of a large number of invading organisms, we do not at present know any means of destroying them without injury to the person whom they have invaded; but, admitting this to the fullest extent, we believe that we shall carry with us the assent of every experienced clinical physician when we affirm the power of judicious treatment, even of drug treatment, to control or modify many of the results of what Dr. Wright calls the conflict.[72]

Another account praised Wright's researches into the opsonic index as "medicine of the future" but still warned that "we are somewhat doubtful, however, about the necessity of providing for "scientific skilled direction of medical "research." Human nature being what it is, such direction tends to degenerate into obstructiveness and fossilism."[73] Wright's claim that the opsonic index could revolutionize medical therapy had clearly met with opposition from clinical physicians.

Even though Wright was not lacking in medical critics, members of the biometrical school seized upon his method of determining the opsonic index by taking averages, to argue for the usefulness of mathematical statistical methods. They were just as determined to create statistical inference as a category of professional expertise as Wright was to establish the importance of bacteriological research. The subsequent debates between Wright and the biometricians even helped to launch the career of one of Karl Pearson's medical students, Major Greenwood (1880–1949).

Greenwood could be seen as one of the first individuals to respond to Pearson's "crying need" for the medical profession to become cognizant of statistical methods. At the age of 18, he entered medical school and read Pearson's *Grammar of Science*, a work he later claimed "did for me what it has done for other lads and I found my intellectual interest."[74] He wrote to Pearson initially in March of 1902 while still a student at the London Hospital, describing himself as an "amateur mathematician" who believed that "some very important and interesting conclusions might be drawn from the material of the Pathological Departments of our hospitals" on the coefficients of correlation between the weights of the spleen, kidneys, heart, liver, and brain in normal and diseased individuals. Greenwood specifically inquired of Pearson if such research would be of scientific interest and if he could conduct this project under Pearson's tutelage.[75]

Two years after this initial inquiry, Greenwood obtained his license to practice medicine, published an article in *Biometrika*, and elected to study under Pearson during the academic year 1904–1905 in order to improve his mathematical skill. His decision was intimately linked to a desire that he "devote [himself] . . . entirely to pure science" because he could "imagine no keener pleasure than research for its own sake." Although Pearson agreed to instruct Greenwood, he warned him:

> It is very difficult with biometry [to pursue a career]; there are at present no teaching posts, no demonstratorships, or fellowships to aid a young man on his way. There will be no doubt one day a demand for statistically-trained medical men for registrars, officers of health, and even later as writers of medical textbooks . . . *But* they may have to wait . . . & perhaps to teach the public first, that the present holders of such offices are incompetent men, and this is a thankless task, and the prizes may come too late.[76]

Despite this warning about the difficulty of earning a living as a biometrician, Greenwood resolved "to be an honest soldier under your [Pearson's] flag."[77] Greenwood's decision to pursue a life of scientific research was solidified in 1905 when he was given a research scholarship by the British Medical Association and became the demonstrator in the physiological laboratory of Leonard Hill in the London Hospital Medical School, working on caisson disease, a position he held until January 1910.[78] Greenwood had staked his professional career on the application of mathematical statistical methods to medical problems.

The first issue that Greenwood chose to address in his new social role as a medical statistician was Almroth Wright's research into the opsonic index. The principal criticism that Greenwood and the biometrical school were to level against Wright was that the experimental method of the laboratory was no longer sufficient to render scientifically valid results; the use of samples implicitly raised the problem of statistical inference. As Greenwood observed to Pearson in 1908:

> The fact of the matter is that we are standing at the parting of the ways in medicine. The day is gone by when *purely* experimental work in either physiology or pathology can greatly advance knowledge. In the days of Ludwig, Claude Bernard and Pasteur the field was comparatively open; it is so no longer and the need for more rigorous logic and statistical methods of analysis must be realised sooner or later even by the average consultant. The experimental method which once saved medicine will ruin it yet. With all your experience you have no conception of the intense badness of much that appears in the medical press under the name of experimental science.[79]

It was from this perspective of showing that "scientific" conclusions had to be based not only on correct experimental procedure but also on valid inferences from experimental data, that Greenwood analyzed Wright's results in the 1908 volume of *The Practitioner*. He addressed the issue of the extent to which the "opsonic index" could be trusted as a guide in the diagnosis and treatment of disease. After determining the mean opsonic index, Greenwood plotted the frequency distribution for a series of measurements and found them to be markedly asymmetrical or "skew." He saw this as a serious problem for Wright's attempt to make diagnoses on the basis of the opsonic index, because the "whole validity of the ordinary 'probable error' method depends upon the assumption that positive and negative deviations are equally frequent, and are distributed symmetrically on either side of the mean value."[80]

Greenwood continued his critique of the trustworthiness of the opsonic index in *Biometrika*. He maintained that the question of the accuracy of the predictive value of the opsonic index was "essentially a statistical one" and that "it is statistically complex and cannot be solved without such a preliminary analysis of the data as has, up to the present, not been attempted."[81] In a further act of disciplinary self-justification for the statistician, Greenwood stated, "Since the fundamental point is whether and how far the character of the cells counted (a sample liable to the full error of random sampling) describes fairly that of the whole population of cells from which it is drawn, the question can only be adequately treated by a statistician."[82] With his colleague J.D.C. White, Greenwood then reiterated their basic criticisms of Wright's method of cell counting from a statistical standpoint. Since pronouncedly skew distributions were produced, the various statistical tests based on the assumption of a "normal" distribution could not be readily applied.[83]

Greenwood did not limit his discussion of the function of the professional statistician in medicine to house journals like *Biometrika* where he was largely preaching to the initiate. He also argued before the pathological section of the Royal Society of Medicine that there was a unique role for the professional statistician. He invoked a distinction that would serve as a central point of contention between the statistician and the medical researcher, the difference between functional and mathematical error. The former concerned errors of technique that could result in incorrect facts being recorded, while the latter concerned inferential errors derived from the fact that conclusions were based on a *sample* of the population rather than the group as a whole. Errors of the former type were seen as the province of the medical researcher, and errors of the latter type, as the province of the statistician.[84]

Speaking as a professional statistician, Greenwood arrived at three conclusions regarding the "mathematical" errors in Wright's data. First, he observed that the frequency distributions produced were markedly asymmetrical, i. e., positive and negative distributions from the mean did not occur equally often in random sampling. Second, he noted that this asymmetry was reduced, but not removed, by emulsions that had maximal thickness (i.e., a larger number of bacilli are found in each cell). Third and finally, Greenwood observed that the mode or most frequently occurring value was a more accurate constant on which to base analyses than the mean as a result of the asymmetry of the distribution.[85] In a self-conscious act of disciplinary justification, he proclaimed: "This must be almost the first time a statistician has had the temerity to address you on a definitely pathological subject. The time will come when such co-operative work is not the exception but the rule, to the benefit of both kinds of investigator."[86] The subsequent commentators on Greenwood's paper praised his "lucid presentation of the statistical side of the question" but inquired if there was an "easy method" of determining the mode rather than the mean. In the report on the ensuing discussion, it was noted that Greenwood "did not see any satisfactory way of determining the mode simply."[87] Although impressed by Greenwood's technical competence, the physicians questioned the practical value of his statistical arguments .

Greenwood was not alone among members of the biometrical

school in using the opsonic index to argue for the joint effort of the
professional mathematical statistician and the professional bacteriologist in the domain of medical research. The issue was nicely
framed in *Biometrika* by W. F. Harvey, another of Pearson's disciples:

> Much misunderstanding seems to exist as to the relationship which
> should hold between the statistician, anxious to scrutinise the validity of
> inferences drawn from observations, and the observer of the facts. The
> latter looks with suspicion on the former and is inclined to doubt
> whether the intrusion of his fellow scientist into his domain is for any
> more worthy purpose than simply to show that he is totally wrong in his
> conclusions. The mathematician is apt to be impatient at the want of
> knowledge of his technical terms displayed by the observer and at the
> often openly expressed contempt for some of his most cherished ele-
> mentary principles. But the statistical mathematician must get his data
> from the doctor—for it is of medical science in particular that we speak
> here—and the doctor at present is dependent on or would be wise to
> obtain the assistance of the mathematician for the refining of his con-
> clusions. The doctor may also obtain much help from the criticism of
> the mathematician regarding the numerical soundness of his control
> observations.[88]

Wright was not interested in the attempt by Greenwood and others
to create a professional role for the mathematical statistician in the
field of medical research. Rather, his principal concern was to establish the importance of bacteriological and immunological methods in
the face of skepticism from already-established branches of the medical profession such as clinicians. In a 1910 speech before the Royal
Society of Medicine entitled "Vaccine Therapy: its Administration,
Value, and Limitations," Wright criticized the clinician for an unwill-
ingness to learn modern bacteriological methods. He claimed that
the clinician "is still bemused with the idea that the final appeal must
always be to himself and to his methods of physical diagnosis."[89] For
Wright, the advances in bacteriology in general (and vaccine therapy
in particular) required that even clinicians become knowledgeable
about bacteriology:

> If we, as a profession, deprecate treatment by the unqualified on the
> ground of the dangers which may attach to the treatment of grave cases
> by the ignorant, can we then refrain from condemning, as perilous to
> the patient, the treatment of grave bacterial infections by those who are
> ignorant of bacteriology? And if we as a profession condemn consulta-
> tion by correspondence on the ground that a trustworthy opinion can-
> not be based upon medical data which are furnished by an ignorant

patient, how then shall we refrain from condemning the system by which a medical man who is ignorant of bacteriology selects the bacteriological data upon which a diagnosis is to be based?[90]

Wright stated his views before the Royal Society of Medicine because the medical profession viewed vaccine therapy as a controversial procedure. His views sparked debate at six meetings from May 23 to June 22, 1910, with thirty-three individuals contributing, who reflected the concerns of both clinicians and bacteriologists. The clinicians generally rejected Wright's contention that they should learn about bacteriological methods. They saw the empirical judgment of the clinician and the scientific knowledge of the bacteriologist as playing distinct, though mutually reinforcing, roles in the determination of therapy. As one speaker before the Royal Society of Medicine summarized the issue: "The clinician will remain captain at the bedside. Similarly, the bacteriologist will be captain in the laboratory. The best results of vaccine therapy, or indeed of any application in clinical medicine of the allied sciences, must come from the properly coordinated work of the clinician and the laboratory worker."[91]

Several speakers before the Royal Society of Medicine openly acknowledged that the efficacy of vaccine therapy could not be conclusively established from the already-collected statistical data. They believed, moreover, that the medical profession itself could interpret the meaning of the statistics; they did not appeal to the "expert" opinions of Greenwood or Pearson. T. J. Horder (1871–1955) maintained that until all the patients in one hospital suffering from a disease allegedly treatable by vaccine therapy were allowed to act as controls (by not receiving therapy) while all the patients suffering from the same disease in another hospital were given therapy, "the value of vaccine therapy can never be put upon a sound scientific basis."[92] Similarly, the bacteriologist William Bulloch (1868–1941) observed:

> In estimating the value of any therapeutic method the elimination of chance is of fundamental importance. There should be parallel series of cases, those with and those without the treatment in question; and it is only when this is done on a big scale that the true value of vaccine therapy will be known. At present we do not appear to possess accurate data of this kind, and thus we have to rely only on the personal impressions created in the minds of medical men who have witnessed the treatment of infective disease by different methods.[93]

Although Bulloch had clearly recognized that there was an element of what Greenwood would have chosen to call "mathematical error" in Wright's results, he did not see this as a problem requiring the expertise of the professional statistician (as Greenwood would have desired). Rather, Bulloch saw the dispute as internal to medicine; it involved only clinicians and bacteriologists. Bulloch summarized the issue as follows:

> The profession can be resolved into four groups with regard to the question of the value of vaccine therapy. There is, first, the clinician, who boldly and publicly asserts that as a result of his experience he is unable to subscribe to the view that the results of vaccine therapy are better than those obtained with other well-tried methods. This group appears to be small. There is a second much larger group, who allege this in private, but, for various reasons, do not assert it in public. This seems to include a large number of clinicians. A third and small group condemn vaccine therapy by word of mouth, but practice it by hand.... Lastly, there is a group, in which I reckon myself, who consider that results can be achieved by vaccines which have not hitherto been obtained by other and older methods.[94]

In light of such vehement criticism from clinicians, Wright and his bacteriological associates did not wish to engage the biometricians also. Thus, Wright downplayed Greenwood's distinction between "mathematical" and "functional" errors by appealing (like the physiological critics of Radicke) to the informed-professional judgment of the bacteriological researcher:

> I have satisfied myself, and all my fellow-workers have satisfied themselves, and I am glad to say a very large and increasing number of bacteriological workers all over the world have satisfied themselves, that when the "functional error" has been reduced, as it can be by practice and patience, to small dimensions, and when, in connexion with tubercle, the customary counts of 100 or more leucocytes are made, the "mathematical limit of error" of the opsonic index is such as need not seriously be taken into account.[95]

This view that the mathematical considerations of the statisticians were not serious concerns for the trained bacteriologist was echoed by one of Wright's associates, Leonard Colebrook, who maintained:

> Sir Almroth Wright has not touched to-day upon its [the opsonic index's] accuracy, and I must leave it to someone else to bring forward the more subtle mathematical considerations with reference to it. But with many of us, I fancy, these more refined data and arguments will weigh less than such evidence as the following, which has evolved from our routine blood examinations of cases in the wards and out-patient

department, quite apart from any definite investigations upon this point; treatment being the only question at stake when the "bloods" were tested.[96]

Wright's lack of concern with "mathematical error" was obviously not acceptable to Pearson and Greenwood; they were just as concerned with establishing statistical inference as a category of professional expertise as Wright was with establishing the importance of laboratory methods. When Wright referred to "functional error," Pearson underlined the term in his copy of the paper and wrote in the margin, "Speaks as if he knew what this amounted to."[97] Also, Pearson wrote to Greenwood: "It [the debate about the opsonic index] will be a big fight, but I think we shall ultimately be successful. What passes as proof now in medicine is unworthy of the name of scientific reasoning."[98]

Karl Pearson made known his criticisms of Wright to his followers in an article in *Biometrika* entitled "The Opsonic Index—'Mathematical Error and Functional Error.'" Arguing for his own professional competence in matters of statistical methodology, Pearson declared that until Wright had submitted his results to the test of "trained statisticians," he should not be surprised that the statistical community regarded his claims with skepticism.[99] The problem of statistical inference was as much a professional problem to be handled by the statistician as the changes in the blood were a professional problem for the bacteriologist:

> I have no prejudice for or against the opsonic index method; for me it is a nice problem in statistics, that is all; but I should much like to see a count of 1000 leucocytes made on a Wright Laboratory slide by one of his staff, and then made on the same slide by an independent microscopist not trained in Sir Almroth Wright's Laboratory. I hardly think, as at present advised, that there would be an appreciable difference in the result. Until this be done, it is scarcely scientific—without publishing evidence of any kind—to appeal vaguely to the "satisfaction" of "a large and increasing number of bacteriological workers all over the world." Statistics on the table, please! I may be quite in error, but at any rate the evidence on which my conclusions are based is here provided for criticism and correction.[100]

Greenwood concurred. As he wrote to Pearson in November of 1912:

> I really cannot see any defense of Wright's criterion of experimental judgment which cannot be urged in favour of a theological "proof." How can Wright, on his own showing, refuse to accept the Athanasian creed, as scientifically proved? The theologians have constantly urged that its

clauses are probative to minds having the necessary spiritual "technic" and experience. In fact the criterion that a scientific law is valid for all normal minds is expressly repudiated. I cannot escape the conclusion that Wright is an enemy not merely of biometry but of all science whatsoever.[101]

Greenwood had staked his professional career on the claim that medical inference could be formalized by statistical methods. He did not wish to see Wright dismiss his area of expertise as unnecessary.

Greenwood did find an advocate for his statistical approach to the opsonic index in the person of Charles James Martin (1866–1955), who had become the director of the Lister Institute for Preventive Medicine in 1903. Martin's support resulted in what Greenwood characterized as an "acrimonious dispute between Martin and Wright" at a private meeting of the Medical Research Club.[102] In 1909 Martin created a statistical department for Greenwood to head at the Lister Institute. Greenwood would later characterize his statistical department as

an inferior copy of the magnificent laboratory of my old master Professor Karl Pearson. Its activities are confined to the application of the methods of modern mathematical statistics to the problems of epidemiology and pathology, we do not propose to trespass upon the fields so efficiently covered by the University College department, i.e., heredity, eugenics and pure mathematical statistics.[103]

In the medical press of the day, the uniqueness of Greenwood's statistical department was heralded. As an article entitled "Mathematics and Medicine" in the *British Medical Journal* observed, "With the exception of the Lister Institute, we are not acquainted with any home of advanced medical study and research which makes provision even for instruction in mathematical statistics."[104]

Greenwood could now criticize Wright for eschewing biometrical procedures not merely as one who saw the potential application of mathematical methods to medical problems (both Gavarret and Radicke had done this), but as one whose social role was to make such applications. His arguments were explicitly framed as *ex cathedra* statements from a professional medical statistician. Following Pearson's advice to respond to Wright in the pages of *The Lancet* where "it really will be read by the medical world,"[105] Greenwood declared:

Sir Almroth Wright has satisfied himself that the application of statistical or biometric processes to medical problems cannot yield trustworthy results. The number of biometrically trained medical men is still small, and, so far as I am aware, I am the only one in this country holding a post expressly created to further the application of biometric methods in medicine and pathology. I seem, therefore, honourably obliged either to acknowledge the justice of Sir Almroth Wright's conclusions or publicly to state the reasons which lead me to think them unsound, and to submit for criticism the principles I hold to be just.[106]

Greenwood savaged Wright for arrogating to himself the right to make inferences from his statistical data regarding the opsonic index. Statistical inference was the province of the professional statistician and Wright, as a bacteriologist, had no legitimate claim to authority in this field. Greenwood observed:

> The biometrician, *qua* biometrician, *discloses the data upon which his conclusions rest, the methods by which he has analysed them, and the conclusions which, in his opinion, may be drawn* [italics in original]. He does not appeal to the experience of statistical experts, nor found conclusions upon a (practical) consensus of (qualified) opinion. It is open to any reader to study his paper or not to study it, to decipher the hieroglyphics or not to decipher them. To gibe at the assumption of authority by statisticians because one has not troubled to master their notation would be as reasonable as to charge v. Kries with attempting to set up as a physiological dictator, and to promulgate dogmas because he writes in German, if the critic does not understand that language.[107]

Greenwood held that Wright's rejection of the expertise of the biometrician was untenable for a variety of reasons. Wright himself had professed ignorance of statistical methods; therefore, he was not equipped to pass judgment on the work of a biometrician. Greenwood also maintained that "[t]he claim of a medical expert that he alone is competent to say what his experiments prove is not merely unscientific, but anti-scientific." He declared that it was the "essence of science" to disclose both the data upon which a conclusion was based and the methods by which the conclusion was obtained. Since Greenwood believed that he had fulfilled these conditions in his analysis of Wright's data, he regarded the bacteriologist's charges as "unjust."[108]

Greenwood's statistical criticism notwithstanding, Wright's method of diagnosis by means of the opsonic index was eventually rejected on more practical medical grounds. It was a very difficult procedure to carry out and the results were often inconclusive. As William Osler observed in the seventh edition of his classic textbook *The Principles and Practice of Medicine* (1909): "The method of Wright and Douglas has been extensively tested during the past three years, and it is difficult yet to arrive at positive conclusions. The independent work in England and in America appears to be against the opsonic index as a trustworthy guide."[109]

Ultimately, the disagreement between Greenwood and Wright highlighted the diversity of responses unique to early-twentieth-century British medicine on how the medical profession should acquire "scientific" credentials. Both Greenwood and Wright held that medical research ought to be predicated on a scientific methodology rather than mere clinical judgment; however, they disagreed profoundly over what constituted "scientific" reasoning. Greenwood's vision of "science" was essentially the method of statistical correlation pioneered by Galton and Pearson. Wright's vision of "science" was laboratory-based experimentation. In a lecture published in 1944, three years before his death, Wright continued to warn against the vaunted claims made by the statistician. He declared that "a perfect statistical experiment cannot be complied with" in actual medical practice because of such considerations as the difficulty of securing a statistically significant control group during an epidemic, when all desire to be vaccinated.[110] In contrast to such a method of statistical experimentation, Wright advocated what he called crucial experimentation to determine, in the laboratory, the causal mechanisms that account for the efficacy of various therapeutic agents.[111]

The contrasting visions of science represented by the bacteriologist (the laboratory) and the biometrician (statistics) was recognized even within contemporary British literary culture and commented on by no less a figure than George Bernard Shaw. In his introductory essay to "The Doctor's Dilemma," Shaw delighted in criticizing both Wright and the biometricians. He noted the economic difficulties of making the opsonic test a standard procedure:

> [T]hough a few doctors have now learnt the danger of inoculating without any reference to the patient's "opsonic index" at the moment of inoculation, and though those other doctors who are denouncing the danger as imaginary and opsonin as a craze or a fad, obviously do so because it involves an operation which they have neither the means nor the knowledge to perform, there is still no grasp of the economic change in the situation. They have never been warned that the practicability of any method of extirpating disease depends not only on its efficacy, but on its cost.[112]

As for the biometricians, Shaw observed that "the mathematicians whose correlations would fill a Newton with admiration, may, in collecting and accepting data and drawing conclusions from them, fall into quite crude errors."[113]

The debate between Greenwood and Wright was similar in struc-

ture to earlier discussions, such as the debates between Risueño
d'Amador and Gavarret or Radicke and Vierordt, but profoundly different otherwise, in that both participants recognized the theoretical
"correctness" of the mathematical methods pioneered by the biometrical school; Greenwood and Wright differed, however, on the
question of whether these methods should be applied to the particular medical issues at hand. With the "autonomy of statistical law" in
the work of Galton, statistics could develop as a professional scientific subdiscipline replete with textbooks and formal training in statistical methods. Pearson provided the institutional structure for this
development in his biometrical laboratory at University College. By
training Greenwood, Pearson had helped to foster the vision of the
medical statistician as a hybrid social being, i.e., a researcher who
understands both medical results and statistical methods. In debating Wright, Greenwood was attempting to carve out a place for the
medical statistician in the world of medical research. At the time of
the debate, he was the only person holding a chair created expressly
to deal with medical subjects biometrically. The attempt by Greenwood and other biometricians to generate a wider audience for such
methods in the medical profession is the next subject of this study.

Chapter Six

THE BIRTH OF THE MODERN CLINICAL TRIAL: THE CENTRAL ROLE OF THE MEDICAL RESEARCH COUNCIL

MAJOR GREENWOOD CONTINUED to proselytize for statistical methods in medicine during his tenure at the Lister Institute. His critique of Almroth Wright in *The Lancet* concluded with an appeal that more of the medical profession become knowledgeable about modern statistical methods, because such methods would be "within the competence of all post-graduate medical workers." Introducing a comparison from the game of chess, Greenwood observed:

> If anyone arrogates to himself the rank of a chess master and proclaims that in any given position the right move is such and such, so many people have a fair knowledge of the game that he will soon be found out if he is a pretender. Unfortunately, this is not yet the case in medical statistics, and too much work is either ignorantly proclaimed to be a "mathematical proof" of something or ignorantly dismissed as "futile trifling." The cure for this state of affairs is to learn the moves. I rejoice to think that the cure is likely to be adopted, and it is pleasing for the advocates of medico-statistical methods to read the tribute of Sir William Osler paid in words which may fitly conclude this paper: "Karl Pearson's new iatro-mathematical school of medicine has done good work in making the profession more careful about its facts as well as its figures."[1]

Greenwood then enlisted the help of his friend and fellow statistician at Cambridge University George Udny Yule (1871–1951) to argue for the usefulness of statistical methods in adjudicating bacteriological disputes. Using a test developed by Pearson in a 1900 paper, they analyzed antityphoid and anticholera inoculation procedures from a statistical standpoint to see if they were truly efficacious.[2] In a paper published in the *Journal of Hygiene*, they argued further that the statistician had special mathematical knowledge that could help in the bacteriological analysis of water. Citing a study that had analyzed the bacilli found in cubic centimeters of water, they observed:

> We think, indeed, that the tenor of the passages cited creates a presumption that the authors' criterion really is that sources *shown by other methods or found from practical experience* [italics in original] to be safe

or to be unsafe have *usually* been found to give sterile readings when samples of the assigned size have been tested. This would explain, for instance, the lower standard adopted in the case of moorland waters. This is undoubtedly a reasonable attitude of mind enough, but it is necessary to remark that the process is not wholly satisfactory, since two observers both testing the same source on, say, the basis of a sample of 100 c.c. might obtain the one a positive, the other a negative result, so that the one would reject and the other pass the supply. Further, no criterion is provided of the increase in accuracy of prediction attained when two, three or more samples of 100 c.c. all give sterile readings.[3]

They proposed criteria to make possible accurate results "in any given case." Yule and Greenwood felt that statistical methods provided objectivity and indisputable results.

In addition to demonstrating in the medical press how biometrical methods could be used, Greenwood used his position at the Lister Institute to further the cause of the professional medical statistician. As he reported to Pearson in 1911, he had begun to offer a course of instruction in statistical methods and numbered among his pupils "the University Professor of Protozoology, the Director of the Institute and other estimable persons."[4] Later Greenwood observed to Pearson that "we even have the nucleus of a fair library."[5]

Despite these early signs of institutional support, Greenwood realized that the use of statistical or biometrical methods in medicine was an uphill fight. As he observed to Pearson, "Immense as is the value of the conception implicit in the eugenic movement, the wider generalisation that the spirit of mathematics must inform *all* science is still more important. I shall not live to see medicine quantified, but perhaps my baby will; in the meanwhile we must fight hard for the ideal."[6] Although Greenwood would later maintain (mixing his metaphors with abandon) that the "inoculation of medicine with logic" was an end that was "in sight with the help of an only moderately powerful telescope," he could still lament the fact that the "mathematical olympians" remained uninterested in medical problems.[7] As a mathematically trained medical statistician, Greenwood still played a unique professional role in the second decade of the twentieth century.

Greenwood did have an American counterpart, however, in the person of Raymond Pearl (1879–1940). Pearl had become interested in biology as an undergraduate at Dartmouth and eventually received a Ph.D. in the subject from the University of Michigan in 1902 by writing a doctoral thesis on the reactions and behavior of planarians. At this time he became intensely interested in biostatistical methods, especially correlation, and wrote to Pearson seeking advice on these subjects.[8] Subsequently, Pearl went to study under Pearson for the academic year 1905–1906 and was eventually made a coeditor of *Biometrika*.

Like Greenwood, Pearl subsequently pursued a career in biological and medical research rather than eugenics per se. Pearson attributed this behavior to the lack of academic positions devoted to eugenic research. As he observed to Galton:

> In the last four or five years I have had at least 2 or 3 really strong men pass through my hands, but I could not frankly say: "Stick to statistics & throw up medicine or biology because there is someday a prize to be had." I feel sure, however, with a future, such men will naturally turn to the Eugenics work. Only this last winter one of my American students said: "I wish I could go in for eugenics, but my bread & butter lies in doing botanical work. I know that definite posts are available." And that was precisely the case with Raymond Pearl, who has now got the control of an agricultural state breeding station—he was far keener on man than on pigs & poultry, but the public yet has not realized that it needs breeding also![9]

For Pearson "statistics" were primarily linked with eugenics and not as important for biology or medicine. He compiled statistics dealing with these subjects to further his eugenic cause.

For Pearson's students who found careers in these other professions, such as Pearl and Greenwood, their general statistical training proved to be more important than a specifically eugenic outlook. When Pearson summarily dismissed Pearl from the editorship of *Biometrika* in 1910, Pearl remonstrated that "I can see no reasons why a firm conviction of the value of statistical methods in biological work should necessitate that one should subscribe to your views as to the method and nature of inheritance."[10]

Greenwood likewise showed skepticism toward Pearson's eugenic orthodoxy. As he observed to Yule in 1913:

> All this chatter about nutrition having no relation to, not the *Anlage* of intelligence—that is something we know nothing about—but the manifestation of the *Anlage* as shown in the shaping of the child at school either in work or in the impression he produces on the teacher is manifest balderdash. Give a dog a protein-free diet and he will become a corpse after a certain number of days, give him protein but not enough to keep him in nitrogenous equilibrium and he will equally become a corpse in a rather greater number of days. Now we know that many of the kids are not in nitrogenous equilibrium. . . . All this is not just medical dogma but hard solid experimental fact. Really if this is all we statisticians can do towards the solution of social problems . . .[11]

Pearl and Greenwood were willing to spread Pearson's mathematical statistical approach within the medical profession; however, they had now rejected his eugenic outlook.

To further the cause of such a statistical approach, Greenwood and Pearl created careers for themselves, as statisticians, in the medical research establishment. In 1918 Pearl began a long-standing relationship with The Johns Hopkins University as professor of biometry and vital statistics in the School of Hygiene and Public Health (1918–25) and as statistician at The Johns Hopkins Hospital (1919–35).

Greenwood, likewise, solidified his connections with the medical establishment. In 1920 he left the Lister Institute for a position at the Ministry of Health and became affiliated with the newly created Medical Research Council (MRC). The central institute of the Medical Research Committee (a forerunner of the MRC) had been set up in Hampstead in 1914 and became the National Institute for Medical Research (NIMR) in 1920. Initially it consisted of departments of bacteriology, applied physiology, biochemistry and pharmacology, and medical statistics. The medical statistics department was headed at the outset by John Brownlee, who was educated in Pearsonian statistical methods.[12] Greenwood characterized his early relationship with the MRC as "purely topographical and unofficial," i.e., he had rooms in their institute and had friends on the staff, but his principal appointment was at the Ministry of Health.[13]

Greenwood clearly saw his position within the medical establishment as useful for spreading knowledge of statistical methods. When Pearson attempted to have Greenwood appointed as reader in medical statistics at University College, Greenwood balked, declaring that "the chance of doing a little good at & through the Ministry of Health would be lost." He believed that the statistical cause could be advanced more through educating the general medical practitioner than by devoting himself totally to research in the Galton Laboratory. As he observed to Pearson:

> I gather that you regard the teaching function as quite subsidiary to the research function (indeed this seems a necessary inference from the founder's words) and would not welcome undergraduate students not likely to develop into efficient research workers. Now when I leave the Ministry I want to make these rank & file my chief care because I believe that the low standard of general scientific culture, the utter ignorance of what happened in medicine, particularly epidemiology, before Koch discovered the tubercle bacillus, is largely responsible for our present plight. Of course neither I, nor anyone else, can teach at all without some facilities for research and without devoting much of one's energies

to research. But if I had to choose between my research & clever pupils on one side or trying to lighten the darkness of the average medical student on the other, the average man should have the preference.[14]

Pearson was more interested in maintaining the cohesiveness of his research institute than in furthering a biostatistical approach to epidemiology. The public interest in eugenics had begun to wane. The computing skills of the biometrical school were utilized for ballistics research during the war, and several members of Pearson's staff subsequently took jobs in government service; inflation ate away at Pearson's financial resources; and the work of the Eugenics Society markedly declined during the 1920s.[15] Pearson's offer of a position to Greenwood in his laboratory was an attempt to maintain eugenics as an academic specialty. He wrote to Greenwood that "while we have a whole world to convince of the importance of our science—and really it is almost the Novum Organon of all true science—we cannot stick too closely together and pull in the direction of agreement."[16] Eventually an agreement was reached whereby Greenwood would be made Honorary Reader in Medical Statistics with no salary because of his official appointment at the MRC.[17]

Although Greenwood maintained such a *pro forma* relationship with the Galton Laboratory, he was more interested in creating the medical statistician as a professional role within the medical establishment than with furthering the eugenic concerns of his former mentor Karl Pearson. When Pearl advised Leonard Darwin that Greenwood should be appointed to the Galton chair when Pearson stepped down, Greenwood maintained that he "should no more dream of becoming a candidate for it [the Galton chair] than of becoming a candidate for the presidency of the United States." He observed: "The Eugenists are a sorry crowd. Leonard Darwin is a dear old thing but exhibits a steep regression towards mediocrity from the paternal standard of brains and most of the others are mob orators or women who think that heart to heart talks on ways of inducing other women to eschew quinine pessaries are eugenics."[18] He indicated that he was primarily a statistician rather than a eugenicist:

The Galton professor ought to be primarily interested in eugenics not medical statistics and as a matter of fact neither the Pearsonian nor the Leonard Darwinian brand of eugenics interests me at all. The latter seems to me merely after dinner tosh, the former like H. G. Wells's characterisation of the Times newspaper, i.e., someone of the greatest possible importance with a cold in the head talking through layers of felt. In the intensely improbable event of a complete failure of superficially eligible candidates and a consequent direct offer to me, I would at least consider the subject. But this is so remote a contingency that it is not worth writing or thinking about. A professorship of statistics sans phrase or of medical statistics would be quite another proposition.[19]

By the early 1920s, Greenwood was not alone in arguing for an approach to medical statistics based on mathematical considerations. Programmatic statements for the adoption of statistical methods became more frequent. As one writer observed in the *Journal of the American Medical Association* in 1920, "The elementary theory of probabilities which deals scientifically with these [statistical] methods is comparatively a simple branch of mathematics, is of great practical significance, and should be required in the premedical curriculum."[20]

Raymond Pearl made a similar proclamation on the potential usefulness of statistical methods in a 1921 article in the *Johns Hopkins Hospital Bulletin*. He began by noting the advances that had been made in statistical methods following the pioneering work of Galton, Weldon, and Pearson at the end of the nineteenth century and the controversies that had surrounded the extension of these methods into the medical domain. Then he advocated the general introduction of statistical methods into medical instruction declaring:

1. That there is no inherent reason why medicine in every one of its phases should not ultimately become in respect of its methods an *exact* science, in the same sense that physics, chemistry, or astronomy are to-day exact sciences.
2. That this goal will be reached in exact ratio to the extent to which quantitative methods of thought and action are made an integral part of work of every sort of medicine.
3. That no number or figure can be said to have any final scientific validity or meaning until we know its probable error, the "probable error" being the measure of the extent to which the number will vary in its value as the results of chance alone.[21]

Pearl concluded that these propositions could be institutionalized by requiring that the quantitative data generated by the modern hospital be analyzed in cooperation with the expert statistician. The arguments for introducing statistical methods into medicine were framed in terms of the broader attempt to ensure that medical research become "scientifically" grounded.

Greenwood shared Pearl's view that statistical methods were a means for medical research to acquire "scientific" foundations. He perceived that the principal problem facing the statistician was to convince the public at large of the fruitfulness of such methods. As he observed to Pearson late in 1924:

> I cannot get young medical men willing to specialise in statistics for the obvious reason that there is no effective economic demand for medical statisticians, the creation of even half a dozen posts (a very unlikely event) in the public service would not create an *adequate* demand; such a demand will only exist when *non-medical* public opinion insists on the employment in municipal services of as many medical statisticians as it now requires bacteriologists and pathologists. I am certain that demand will be made, but not in my life time; medical statistics are now in the position of biochemistry forty years ago.[22]

One method that would-be statistical professionalizers such as Pearl and Greenwood used was the publication of textbooks explaining to a medical audience the "proper" method of statistical reasoning. In his 1923 textbook, Pearl asserted:

> It is evident enough to every thoughtful observer that clinical medicine is proceeding by great strides along the quantitative, scientific pathway. Every step in this direction adds to the necessity of the medical man having at his command the necessary elementary principles for dealing easily, confidently, and accurately with quantitative data.[23]

This textbook was praised among the small community of individuals knowledgeable about biometrical statistical methods: G. Udny Yule, the holder of a chair in statistics at Cambridge, called the work "attractive" and "useful" to students; Greenwood characterized it as "a most valuable piece of work"; Pearson said the work would be of "great value to both instructors and students in medical statistics."[24]

One of the problems facing Pearl and Greenwood in their attempt to create statistics as a category of professional expertise was the peculiar manner in which statistical methods related to society in general and to science in particular. Medical statistics had been socially and culturally pervasive long before the time of Greenwood or Pearl; anyone who collected aggregative data about a medical subject could be seen as a medical statistician. What Greenwood and Pearl were attempting was to reconstitute the nature of statistical knowledge; it was now to be a technical mathematical discipline replete with all of the characteristics of a scientific profession. There should be textbooks, formal courses of instruction, and peer review of publications.

Other leading members of the medical profession did not share this view that one needed formal *training* in statistical methods. These medical researchers, like their nineteenth-century predecessors, placed little emphasis on mathematical methods when collecting statistics. Greenwood, for example, was critical of Pearl's famous Johns Hopkins colleague William Halsted for failing to take age distribution into account in reporting statistics on the success of operations on cancer of the breast. He observed that "surgeons in this country are mostly at the intellectual level of plumbers, in fact just well paid craftsm[e]n, I should like to shame them out of the comic opera performances which they suppose are statistics of operations and a really decent set of figures from such a panjandrum as Halsted would go a long way."[25]

In addition to indifference from medical colleagues, the mathematical statistician also had to contend with disrespect from "pure" scientists. As a practitioner of a quintessentially applied discipline, the statistician could neither interest academic mathematicians in his or her work nor benefit from the rewards structure of academically based science disciplines. In his correspondence with Pearl from early 1923, Greenwood characterized mathematicians as "psychologically incapable of real co-operation in biological researches" and observed that it was difficult to train statistical epidemiologists because he never got a medical student and "it is rather silly to lecture exclusively to mathematicians who have not the faintest intention of becoming vital statisticians or epidemiologists." Pearl's book was praised as the only source that provided medical students with training in statistical methods.[26]

Pearl concurred with Greenwood that more "traditional" scientists rated the statistician as secondary in importance. In response to Greenwood's statement in 1927 that he might be admitted to the Royal Society, Pearl observed:

> It always seemed to me to be one of those things that did not matter much after you had it, and wanted a lot of explaining if you did not have it. I mean this, of course, relative to persons established in recognized branches of science, like physics, chemistry, etc. Of course a statistician, from the standpoint of other scientific men, is an outcast anyhow, and has to be about six times as good as the ordinary man to get elected to such bodies as the Royal Society. We are having the same sort of difficulty here electing Glover to the National Academy. To be sure, Glover is no great genius, but at the same time he is a better man than a good many of the mathematicians and astronomers that we have lately

elected to the Academy. But the mathematicians in the Academy will have nothing to do with him because he has been tainted and corrupted by statistics.[27]

Despite these protestations that the medical statistician was a kind of second-class citizen in the world of scientific and medical research, the reputations of both Greenwood and Pearl were on the rise at this time. In 1927 Greenwood left the Ministry of Health to become the first professor of Epidemiology and Public Health at the London School of Hygiene. The position had been promised to Greenwood as early as 1923, but it took the school four years to get itself established.[28] With the death of John Brownlee in 1927, Greenwood became the virtual head of the statistical research unit at the MRC. He observed to Pearl, "I can consolidate the little kingdom fairly satisfactorily."[29] In addition, the Rockefeller foundation had begun to fund those who wished to learn statistical methods. As Greenwood wrote to Pearl in early October of 1924:

I wonder whether you and I should not enter into a concordat as to the training of the Rockefeller pups. They seem to me to be let loose to wander through the world in a most casual way. If we were furnished beforehand, say three months before the wanderers started, with a list of those desirous of including statistics in their studies, we could advise the Trustees whether they should go to America or to England. The School of Hygiene does not even contemplate starting work for two years so that as far as England is concerned and medical statistics I am a monopolist; K. P. has not yet broken the truce so that I can ensure a welcome there for the very rare bird fit to gather the correlation ratio from its original tree.[30]

Pearl, likewise, was receiving academic accolades. In 1925, he was named director of the newly created Institute for Biological Research, and his academic title at Johns Hopkins was changed to research professor of biometry and vital statistics in the School of Hygiene. The institute was supported by annual grants from the Rockefeller Foundation and was designed to investigate population growth, duration of life, and the relationship between constitution and disease. After Pearl's appointment, the principal teaching duties for training public health workers in biostatistical methods were taken over by one of his associates, the mathematician Lowell Reed.[31] Pearl was ecstatic that he could now devote his life to research. As he wrote to Yule, "I will be free from teaching and executive duties from this time on and can really get down to work at the thing which interests me more than anything else in the world."[32]

The institutional base provided by these appointments gave Greenwood and Pearl the opportunity to create a new approach to medical research, namely experimental epidemiology. In contrast to the kind of crucial experiment that would interest a Bernard or a Wright, the experimental epidemiologist would focus on the introduction of a disease into a population and then use methods of statistical correlation to determine the effect of factors such as age, sex, or weight on the spread of the disease through the population. In these researches the methods of both the laboratory (to diagnose and treat the disease) and the statistician came into play. The issue was how these two distinct experimental methods should be related, not opposed (as represented by Bernard), in the context of epidemiological research.

Pearl saw in statistical methods a way to make more precise the results obtained from laboratory experimentation. When Greenwood (always more of a statistician than an experimentalist) wrote to Pearl regarding the "simplicity of human lab. experiment," the latter retorted:

> I am compelled to infer that your friend, Almroth Wright, has got you buncoed about the "simplicity" of experimentation. As a matter of fact, it is of the most extraordinary difficulty to set up in a laboratory any experiment which will give anything approaching unequivocal or crucial results. A great many people delude themselves on this point rather tremendously. It is because they either are not experimentalists—that is, lacking what is essentially an artistic flair for this sort of thing,—or else, they are quite unable to think critically. ... I have thought about this question a long time, and the general conclusion to which I have quite undogmatically come is that by and large multiple correlation and laboratory procedure are on pretty much the same footing, taking the whole range of natural phenomena to which either method might be conceivably applied into account.[33]

Greenwood later concurred. Even though he acknowledged the added cost to perform an experiment on a statistically significant population, he still maintained that such a procedure was necessary to make advances in epidemiology. As he wrote to Pearl:

> We have used up thousands of mice and in an experiment designed to test one point only are using terrible numbers. Mice are dear and laboratory servants dearer. But this sort of thing must be done or people will chatter about epidemiology to the end of time. I am not such a fool as to suppose that *when* we know the "law" of pasteurellosis of mice, we shall know the "law" of scarlet fever, dip. etc. but I do hereby proclaim IN

LARGE CAPITALS that until we know the former we cannot begin to know the latter. I am sometimes, perhaps when I am tired and bored, tempted to doubt whether modern statistical developments have not done actually harm by tempting people to suppose that with such sharp tools they really can hack a way to truth through human statistics without leaving their studies or biometric laboratories.[34]

Although not without flaws, statistical methods were a way to bring professional "scientific" legitimacy to epidemiological research.

In addition to carrying out research and training students, Greenwood argued for the importance of mathematical statistical methods by writing historical accounts of medical statistics; he wanted to show how previous historical epochs had fallen into error by not utilizing "modern" statistical methods. As early as 1924, Greenwood wrote to Pearl on the factors that prevented previous ages from appreciating statistical methods. He maintained that earlier times had not perceived the "*psychological* necessity of statistics":

> What prevented the old really clever epidemiologist like Ballonius and Sydenham from really finding out any general truths about the laws of incidence of disease was the individual human interest of their cases. If you start out to write an objective account of the succession of diseases and cannot cloak the individualities of your patients under numbers and percentages you can't help being side tracked by particular experiences. . . . That was why old Graunt, not half so clever a man as Sydenham and no doctor, found out ten times as much as Sydenham did, the human side of disease and suffering did not throw him out of his stride.[35]

Eleven years later, in a textbook on epidemiology, Greenwood praised the "new" statistical methodology of Adolphe Quetelet, Francis Galton, and Karl Pearson. Unlike the earlier French theorists Condorcet and Laplace, the British biometrical school had been more circumspect in its statistical inferences. Greenwood concluded that many mistakes attributed to these later British statisticians "exist only in the minds of their critics."[36]

In his 1936 study *The Medical Dictator*, Greenwood cast Louis and Poisson in the role of visionaries who heralded the modern age in which mathematically based statistical methods guided clinical studies. Observing that Louis did not attempt to determine the reliability of his clinical statistics, Greenwood declared:

> One wonders what might have been the future of clinical statistics if Louis had secured the collaboration of a contemporary who was as faithful a servant of truth and as laborious as himself, S. D. Poisson. If

Poisson's mathematical genius had been applied to the probability of medical diagnoses rather than of the correctness of legal decisions, we should still have had Poisson's "law" and we might have had a French school of clinical statistics which would have led the world.[37]

Greenwood was obviously ignorant of the work of Gavarret. He referred instead to the criticisms that had been leveled against Louis by Risueño d'Amador and others at the Academy of Medicine debate of 1837. He observed that such critics had "assumed that Louis's statistics were valid but denied that they were valuable."[38] As a statistician, Greenwood was more concerned with portraying Louis as "a blood brother of Farr and Galton" and lamented what he called the "real tragedy" that "the systematic use of statistical methods in clinical medicine has never entered into the *routine* of hospital work."[39] Although clearly a historical work, Greenwood's narrative also served as a programmatic statement for the wider dissemination of statistical methods within medicine.

Greenwood's various attempts to create a wider domain of application for medical statistics had begun to have an effect by the early 1930s, at least at University College. The issue was nicely highlighted by the events surrounding the retirement of Karl Pearson in 1933. When Pearson announced his retirement, a proposal was put forward that the Galton laboratory be split into separate departments of Eugenics and Statistics. Pearson did not respond favorably to this suggestion. His statistical approach had been one of the methods he had used to argue that his eugenic research had "scientific" underpinnings. The separation of statistics from eugenics threatened Pearson's professional legitimacy. In an impassioned nineteen-page history of the Galton laboratory, Pearson declared, "The Committee of the professional Board who drew up the scheme were in the bulk new-comers to the College and unfortunately ignorant of the history of the Laboratories, of Galton's plans, or the exact origin of the grants upon which the income of the laboratory has been based."[40] Pearson's protestations proved to be of no avail. The tension between a general belief in the universal applicability of statistical methods and a particular belief in eugenics, embodied in the professional persona of Karl Pearson, could no longer be sustained. The officials at University College concluded:

There is no such essential connection between the study of Statistics and the study of Eugenics as would suggest that the Department of Statistics should necessarily be associated with the Department of Eugenics; their functions are entirely different and a Department of Statistics must stand in as close or loose a relationship to many other subjects of study (for example, Economics and Physics) as to Eugenics.

Statistics is a subject of ever-widening interest; in order that the subject may develop freely it must not be hampered by being linked to one particular field of study in which its methods may be applied.

On these various grounds they recommend that so far as the staff of the Department is concerned there should henceforward be two independent Departments, viz. a Department of Eugenics, including Anthropometry, and a Department of Statistics.[41]

The professional emergence of statistics as a codified body of knowledge and the concomitant rise of individuals trained in its methods provided the necessary conditions for the Laplacian vision of the probabilistically based clinical trial to come into being. The individual who has often been heralded for first executing a clinical trial in its modern form was one of Greenwood's protégés, Austin Bradford Hill (1897–1991). Hill contracted tuberculosis while being stationed in the Middle East at the outset of World War I. Throughout his long convalescence, he studied for a London external degree by correspondence course in economics. Upon completion of his degree, he was offered a post with the Industrial Fatigue Research Board, a semi-independent body under the auspices of the MRC. This position provided him with his early training in epidemiology and led to his contact with Major Greenwood, also at the MRC. Hill's father, Leonard Hill, had been the physiologist with whom Greenwood had trained in the years 1905–10. When Greenwood moved to the London School of Hygiene and Tropical Medicine in 1927, he took the younger Hill onto his staff. Hill learned statistical methods from Pearson at University College and officially became Reader in Epidemiology and Vital Statistics at the London School of Hygiene and Tropical Medicine in 1933.[42]

By this point in time, the various efforts by Pearson, Greenwood, and Pearl to portray the usefulness of mathematical statistical methods in medicine had begun to have an impact. In 1937 the editors of *The Lancet* asked Hill to write a series of articles on the proper method of applying statistics to medical concerns. These articles were subsequently published in book form as *Principles of Medical Statistics*. In the foreword, the editors declared:

In clinical medicine to-day there is a growing demand for adequate proof of the efficacy of this or that form of treatment. Often proof can come only by means of a collection of records of clinical trials devised on such a scale and in such a form that statistically reliable conclusions can be drawn from them. However great may be our aversion to figures, we cannot escape the conclusion that the solution of most of the problems of clinical or preventive medicine must ultimately depend on them.[43]

Hill's book was perceived as necessary because "few medical men have been trained to interpret figures or to analyse and test their meaning by even an elementary statistical technique."[44] Hill took on his essentially pedagogical role for the medical profession with great zeal. He maintained in the preface that "the worker in medical problems, in the field of clinical as well as preventive medicine, must *himself* know something of statistical technique, both in experimental arrangements and in the interpretation of figures."[45] The book proved to be a resounding success. The ninth edition was printed in 1971 and in the twenty-year period from 1962 to 1982 the work was cited nearly eight hundred times.[46]

Despite the success of his textbook on medical statistics, Hill's fame rested largely on the randomized clinical trial that he designed for the Medical Research Council in 1946 to study the effect of streptomycin on tuberculosis. The innovations that made this clinical trial unique derived from developments both internal to statistical theory as well as reorganizations within the MRC as a research institution.

The major theoretical innovation traditionally considered to have spawned the modern clinical trial is R. A. Fisher's work *The Design of Experiments* (1935). In this work Fisher put forward the central importance of the idea of randomization. He developed his ideas in the context of agricultural experimentation comparing the yields of different varieties of grain. Fisher proposed that the experimental plots be divided into narrow strips and that grains be assigned to their place in the field by the use of a chance mechanism (randomization). For Fisher the introduction of chance (and uncertainty) guaranteed that differences in yield reflected a true difference in grain productivity. The personal biases of the individual experimenter regarding grain productivity or a new medical therapy were eliminated by employing the mechanism of chance, which guaranteed that results possessed a degree of "objectivity."[47]

The implications of Fisher's idea of randomization for medical research were recognized by Greenwood, who characterized Fisher's ideas as "epoch-making" and "applicable far beyond the bounds of soil research" in an article published in 1948, the year before Greenwood's death. Greenwood claimed that pioneering ideas like those of Fisher had made statisticians "capable of reaching results in fields which even so recently as 50 years ago seemed closed to them" and that these results contributed to the increasing "demand for statisticians." Greenwood compared this situation favorably to the early twentieth century when Almroth Wright viewed the statistician as "an inferior race of mankind."[48]

In addition to these changes within statistical theory brought about by Fisher's work, changes within the organization of the MRC also facilitated the emergence of the probabilistically based clinical trial. In 1926 the MRC set up a Chemotherapy Committee jointly with the Department of Scientific and Industrial Research, in order to encourage cooperation among chemists, biologists and pathologists, and clinicians in the production of new compounds, in their experimental trials, and in the observations of their effects on human disease. The committee provided fresh impetus to the MRC's involvement in the production and biological testing of new synthetic substances of potential therapeutic significance. New synthetic compounds, many of them prepared at the NIMR, were put forward for biological study or for clinical trial. Other MRC committees were also expected to bring to notice biological products that required clinical appraisal "under controlled conditions." In 1931 the MRC set up regular machinery for the organization of clinical trials by forming the Therapeutic Trials Committee (TTC) in specific response to a request from the Association of British Chemical Manufacturers for an authoritative body to test new therapies. In the trials performed by the TTC, expert opinion was frequently sought from the MRC's Statistical Committee, which was headed by Greenwood. Upon Greenwood's retirement in 1945, Hill took his place both as director of what became the Statistical Research Unit and as professor of medical statistics at the University of London. Under Hill's direction, in 1946 the MRC designed the aforementioned trial to determine the effect of streptomycin on tuberculosis.[49]

The description of this trial, reported in the *British Medical Journal* in 1948, emphasized the meticulous attention to methodology in the study, calling it "a rigorously planned investigation with concurrent controls."[50] To minimize the difference between patients, the designers of the trial imposed restrictions: all patients must suffer from the same disease for which the only known treatment was bed rest; the patients must be between the ages of 15 and 30. As a result of these requirements, the trial was limited to patients suffering from acute progressive bilateral pulmonary tuberculosis of recent origin.[51]

Eventually 107 patients were admitted, with 55 allocated to the streptomycin group and the remaining 52 allocated to the control group with bed rest. Hill used statistical series based on random sampling numbers in order to determine which patient to assign to each group; the details of the series were unknown to the investigators administering the trial, in order to ensure confidentiality. Patients in the streptomycin group received 2 grams of the drug by injection every six hours (i.e., 4 injections each day) for the entire duration of the trial, which lasted six months.[52]

Statistical considerations proved crucial not only in the design of the trial at the outset but also in the interpretation of results at the end. Four of the patients receiving streptomycin died (7%) and 14 of the patients in the control group died (27%) during the six months of the trial. The report concluded, "The difference between the two series is statistically significant; the probability of it occurring by chance is less than one in a hundred."[53] Also, the role of statistical reasoning had become intimately connected with the role of precision measuring instruments; the report observed that the difference in average temperature for the patients in each group was not statistically significant even though there was a drop in temperatures of those who received treatment.[54]

Hill's work had outlined the basic structure within which clinical trials would subsequently be conducted. It was to enlist the professional insight of the clinical physician and employ the professional statistician as an "inference expert" to analyze the quantitative clinical data. The confluence of these two separate traditions constituted the *sine qua non* for the emergence of the probabilistically informed clinical trial. The Laplacian vision of the determination of medical therapy on the basis of the calculus of probabilities had finally found its spokesman. The following decade would bear witness to the increasing influence of Hill's ideas on the direction of clinical research.

Chapter Seven

A. BRADFORD HILL AND THE RISE OF THE CLINICAL TRIAL

HILLS TRIAL was recognized almost immediately on both sides of the Atlantic as the exemplary model of how to conduct an "objective" clinical trial by using randomization. In an early account of the trial in the *Bulletin of the Johns Hopkins Hospital*, the authors observed:

> The report [of the trial] merits study not only for the results but for the way the experiment was conducted. The first part of the report takes up a careful definition of the type of case accepted for study, how the cases were obtained, and how they were allotted to the streptomycin and control groups. ... It is possible to find out from the report just what was done, and what the outcome was for patients falling into various diagnostic classes. The result is that a rather limited number of cases, only 107 all told, have served to give definitive results which one can interpret with confidence.[1]

The authors contended that trials like the one designed by Hill were increasingly necessary because the "large number of new drugs which have been put forward in recent years as agents in the treatment of disease has raised many problems, none of which is more difficult of solution than the decision concerning their safety and efficacy in the treatment of the diseased patient."[2]

From the outset the advocates of the modern clinical trial perceived the conflict between the epistemological desire to achieve certain knowledge and the ethical problem of withholding a potentially useful therapy. The issues were nicely framed in two articles on statistics in clinical research that appeared in a collection of scholarly monographs from 1950 devoted to "The Place of Statistical Methods in Biological and Chemical Experimentation." One essay was written by D. D. Reid, also of the London School of Hygiene and Tropical Medicine, and the other was written by Donald Mainland of Dalhousie University. In his article Reid noted that Hill's trial "expresses very effectively the trend of our thinking on this subject in Britain at the moment." Also, he observed that the unavailability of streptomycin on a large scale (and the lack of sufficient financial capital to expend on such an experimental therapy) enabled them to conduct a study with an adequate control group:

> Our genteel poverty thus paid a scientific dividend by quieting any doubts we might have about the ethics of controlled trials of streptomycin in pulmonary tuberculosis. In short, where only a few could be treated, many would have to remain untreated in the normal course of events. It was decided, therefore, to make the best of this situation by running a well-controlled trial of streptomycin in pulmonary tuberculosis.[3]

With their ethical qualms allayed, Reid cited what the MRC researchers saw as the major advantage of applying the principle of randomization to the design of a clinical trial; it removed the "subjective" bias of the clinical researcher:

> A notable feature of this trial was the frank realization by all concerned of the fallibility of human judgment in general and of clinical and radiological judgment in particular. At all stages of the trial, then, precise criteria of diagnosis, progress, and cure were laid down, and all judgments on X-ray findings were made by two or more observers, independently of each other and unbiased by any knowledge of the nature of the treatment given to the patient whose physical status was being assessed.
>
> The principle of the elimination of personal bias is fundamental in all experiment, but it is of particular importance in clinical research. Thus, in the selection of patients for inclusion in either treated or control groups, the final decision was made purely on a chance basis.[4]

On the basis of these procedures, Reid maintained that the effects of streptomycin on tuberculosis could be determined with confidence because "we were reasonably sure that the rain of chance events had fallen equally upon the just and the unjust."[5]

Despite the vaunted claims made on behalf of statistical objectivity, Reid openly acknowledged that there was still a role for clinical judgment (possibly in an attempt to ward off criticism from the clinicians):

> If the experimental plan has been soundly conceived and executed and the results clash with your notions about the eternal verities of clinical medicine, then, like Cromwell, "I beseech you, gentlemen, consider it possible that you may be mistaken." Statistical methods may be no substitute for common sense, but they are often a powerful aid to it.[6]

Mainland's essay addressed similar themes. He declared that "statistical ideas, to be effective, must enter at the very beginning, i.e., in the planning of an investigation." The clinical experimenter should remain cognizant of the effect of both major factors (age, sex, etc.) and minor ones (psychological, environmental) on the treatment of a patient. Also, he highlighted the critical importance of randomization and chance in assessing the results of a trial: "Having made chance operate in the selection of samples, we can, after the experiment, use our knowledge of chance to assess the results, because we know how often various differences in the results would occur by chance, i.e., if there were no difference between the treatments."[7]

Mainland also confronted what he chose to call the moral problem head-on:

> It is the physician's duty to do his best for his patients, and, if he believes that there is some evidence in favor of a certain treatment, he will feel bound to use it. If, however, he is acquainted with the requirements for valid proof, he will often see that what looked like evidence is not evidence at all, and he will feel impelled, for therapeutic reasons, to alter the treatment in one or more patients. Even then, however, it may be possible to use the data obtained up to the time when the treatment has to be changed.[8]

Mainland concluded his account by praising Hill's streptomycin trial "as a model of experimental design in therapeutic trials" and cited Hill's medical statistics textbook as a source "specially recommended for exposition of principles and elementary methods in the medical field."[9] Ultimately, Mainland affirmed that "modern statistical principles are not something that we can take or leave as we wish, for they comprise the logic of the investigator in all fields, including the field of clinical research."[10]

Following in the wake of this early favorable reception of his trial, Hill actively took on the role of spokesman for mathematical statistical methods within a therapeutic context. Like Pearson arguing for the role of statistics in eugenic research, or Greenwood arguing for its uses in epidemiology, Hill would argue for the extension of statistical methods into the clinic as a way to provide "scientific" results. In this process, Hill knew that he was on new terrain. As he observed in a lecture delivered before the Department of Preventive Medicine at Harvard in 1952:

> It is with mixed feeling that I stand, a layman, before the medical faculty of one of the world's most famous universities. . . . Dare the Statistician now pass from the well-tilled (perhaps I ought to say well-drained) fields of public health to those more exclusive upland meadows in which are practised the arts of the clinician—arts that appear to cultivate the indi-

vidual approach and sometimes even an air of infallibility? It is a bold step. In my own country I have been fortunate enough to be able to take it and can, even at the worst, still exclaim, with the poet Henley, "my head is bloody but unbowed." Over a fairly wide expanse of clinical medicine in Great Britain the statistical approach has been accepted as useful; it is being increasingly applied. But here, like Ruth "amid the alien corn" I stand, if not tearfully, at least a trifle fearfully.[11]

Hill then proceeded to outline what he saw as the principal role of the statistician for clinical medical research. He emphasized that the clinician should enlist the assistance of a professional statistician while the trial was still in the design phase (rather than calling in the statistician after the data had been collected):

The art of applied statistics is, needless to say, compounded of two things—a knowledge of statistical methodology and a wide and detailed knowledge of the data to which that methodology is to be applied. . . . In short, the statistically designed clinical trial is above all a work of collaboration between the clinician and statistician, and that collaboration must prevail from start to finish.[12]

In the speech Hill attempted to answer the criticisms that were often leveled at the clinical trial: it led to "the replacement of humanistic and clinical values by mathematical formulae," the degradation of patients "from human beings to bricks in a column, dots in a field, or tadpoles in a pool," and "the eventual elimination of the responsibility of the doctor to get the individual back to health." Although Hill acknowledged that the medical profession was responsible for curing the sick and preventing disease, he maintained that experimental medicine had also the third responsibility of advancing human knowledge: "The statistically guided therapeutic trial is not the only means of investigation and experiment, nor indeed is it invariably the best way of advancing knowledge of therapeutics. I commend it to you as *one* way, and I believe, a useful way, of discharging that third responsibility to mankind."[13]

In light of Pearson's and Greenwood's views (presented above, Chapter 6) regarding the virtues of a statistical methodology, Hill's arguments ring familiar. Unlike earlier advocates of a statistical methodology within medicine, however, Hill proved to be more than a voice crying in the medical wilderness. His work became a rallying cry for supporters of therapeutic reform on both sides of the Atlantic.[14] Among the several factors that contributed to this groundswell of support, one of the most significant was the vast proliferation of new and highly potent industrially produced drugs in the postwar era. Supporters argued that the clinical trial both permitted the doctor to abandon the time-consuming process of weighing imponderables and curbed the enthusiasm of physicians for newer and more experimental therapies. As L. J. Witts, Nuffield Professor of Clinical Medicine, Oxford University, observed:

> [T]he Second World War may be regarded as the great divide, after which it was no longer possible for the clinician, however distinguished, to discuss the prognosis and treatment of disease unless his words were supported by figures.
>
> The doctor's work in the past had often consisted in weighing a number of imponderables and coming to a decision. He was conscious of the variation between individuals and had felt it part of his skill to decide what would suit one and not the other. On this account practising clinicians have not always taken kindly to the statistical approach to medical problems in which patients are considered as units in a more or less homogeneous whole and their individual variations as an aspect of the normal distribution curve. It is doubtful if such rapid headway would have been made had it not been for the obvious contribution of statistical methods and planned trials to therapeutics. The accelerating pace of discovery entailed the constant trial of new remedies, whose value could not be left to be determined by the slow processes of time and fashion as in the past.[15]

The first edition of Witts's anthology, *Medical Surveys and Clinical Trials*, had appeared in 1959. In March of the same year, Hill had chaired a conference on controlled clinical trials in Vienna. Although the intended audience of the printed proceedings was international (the conference had been sponsored by the Council for International Organizations of Medical Sciences), Hill and his British followers clearly saw themselves as the national leaders in the design of clinical trials. As the foreword noted:

> The Conference was in itself an experiment. The meeting was a closed one, and only one hundred participants were invited. One national group, the British, was charged with the task of presenting each topic to be studied. In this way, all papers were co-ordinated in London by Professor Bradford Hill so that overlap was avoided and ample time allowed for discussion.[16]

For the conference Hill had assembled a band of close-knit followers. There were seventeen contributors, of which eleven had some affiliation with the MRC, the London School of Hygiene and Tropical Medicine, or one of the London hospitals. The conference was divided into nine sections devoted to such topics as the role of clinical

trials in treating acute infections, pulmonary tuberculosis, rheuma-
toid arthritis, coronary thrombosis, and cancer. Papers were also presented on the ethics of the clinical trial, the proper organization and
design of clinical trials, and the contrasting roles of the physician and
the statistician in carrying out a clinical trial.

In their presentations before the conference, Hill, Witts, and Sir
George Pickering, Regius Professor of Medicine at Oxford, all spoke
about the role of ethical considerations in clinical trials. In response
to the clinician's objection to withholding treatment from the control
group, Hill deployed a *tu quoque* argument. He maintained that the
clinician "is *always* using 'controls,' even though those controls may
consist only of his impressions of his past experience with such patients. He cannot escape measurement, whether he like it or not, and
whether it be accurate or not."[17] In addition Hill maintained that the
experimental approach of the statistician in making comparisons was
not incompatible with the observational approach of the clinician.
Rather, "we should seek to incorporate the highly skilled subjective
clinical judgment. We should seek to incorporate it in such a way that
its wielder cannot be biased for or against a particular method of
treatment or—equally important—cannot, though in fact impartial,
be accused of being biased."[18] On the ethical question of when it
would be appropriate to withhold treatment, Hill cited proposals of
other researchers, one of which stipulated that the only trials worth
considering were those in which the physician would allow himself
or a near relative to be included. However, Hill maintained, "the
problem must be faced afresh with every proposed trial." According
to Hill, this was "the only Golden Rule which I feel able to adopt and
offer."[19] Hill concluded by reiterating his claim that a properly designed clinical trial should be a collaborative effort between the clinician and the statistician from start to finish.

Unlike Hill, Witts had no aversion to introducing rules for designing clinical trials. He maintained that three general principles should
be followed: the trial should be conducted only with the voluntary
consent of both the subject on which the new therapy is applied and
those who attend the subject (nurses, social workers, etc.); the clinical trial should seek some benefit to society which would be unattain-
able by any other method; and a trial should be designed to avoid
unnecessary risk and suffering, whether mental or physical. Witts
also downplayed the concerns of the critics of the clinical trial. In
response to what he called "an exaggerated belief in the uniqueness
of the individual whether as patient or doctor," Witts maintained that
neither patients nor doctors were as unique as they may have wanted
to believe: "Empires may rise and empires may fall, but the dose of

opium remains 1/2 to 3 grains."[20] In response to the charge that the individual patient's welfare would cease to be the dominant concern of the physician, Witts conceded that there was "a conflict of loyalties between the search for truth and the treatment of the individual." However, he declared that such a "conflict of loyalties" within medicine did not originate with the advent of the clinical trial; there was a similar conflict between the teaching of clinical students and the treatment of the patient.

As for those who condemned the clinical trial outright because of its experimental approach, Witts claimed that they "are like those Asians who refuse all post-mortem examinations but enjoy the fruits of Western medicine." Witts concluded that the tension between the obligations toward the individual patient and the attempt to advance scientific knowledge was inherent in the nature of medicine as a profession and could not be avoided by eschewing the clinical trial out of hand:

> The categorical negative seldom gets us far in ethics, and virtue is commonly situated midway between two extremes. . . . I believe . . . that the sick person's right to be cared for by the community carries with it an obligation to contribute to the reduction of sickness and premature death, so far as this is possible without physical or mental distress. And finally I believe that there is no stock formula which the physician engaged in therapeutic trials can use to guide his actions. Instead he must have a tender conscience and be prepared to justify all his actions before his Maker. He should also be ready to defend them at an earlier date in the law courts.[21]

In his comments at the conclusions of the conference, Sir George Pickering echoed Witts's praise of the clinical trial. Although a physician himself, Pickering was critical of the vaunted claims made on behalf of "clinical experience." He declared that clinical experience is "unplanned and haphazard" and that physicians "are victims of the freaks of chance."[22] To overcome this uncertainty in the determination of therapy, Pickering advocated the use of the clinical trial. In response to the charge that the clinical trial constituted experimentation on patients, Pickering responded:

> [A]ll therapy is experimentation. Because what in fact we are doing is to alter one of the conditions, or perhaps more than one, under which our patient lives. This is the very nature of an experiment, because an experiment is a controlled observation in which one alters one or more variables at a time to try to see what happens. The difference between haphazard therapy and a controlled clinical trial is that in haphazard

therapy we carry out the experiments without design on our patients, and therefore our experiments are bad experiments from which it is impossible to learn. The controlled clinical trial means merely introducing the ordinary, accepted criteria of a good scientific experiment.[23]

With the emergence of anthologies devoted to the clinical trial like those edited by Hill and Witts, the procedure clearly moved toward becoming a standard for adjudicating issues of therapeutic uncertainty. For the first time, the foundations had been laid for the widespread dissemination of statistical methods within a therapeutic (as opposed to a public health) context; all that was needed was a catalyst to bring the issue of statistical methods into the open.

As with the earlier instances of quantification in a therapeutic context, the initial inspiration came from sources other than the medical mainstream—this time, however, not from lone individuals (such as Gavarret, Radicke, or Greenwood) but from the public at large. Ultimately, the triumph of clinical trial as a standard procedure was due less to internal debate within the medical profession than the emerging belief within society at large that the medical profession and its therapeutic decisions had to be regulated.[24] In this sense the modern clinical trial could be seen as "socially constructed."[25]

Even though Great Britain had taken the lead in establishing the foundations of the modern clinical trial, the United States was the first of the major industrial democracies to institutionalize the procedure as the standard method for determining drug efficacy. The specific issue that established the clinical trial as indispensable in medicine in the United States was the public protest against the use of the drug Thalidomide in the early 1960s. The drug, developed in West Germany, caused an outbreak of infantile deformity and the U.S. Food and Drug Administration (FDA) subsequently discovered that over two and a half million tablets had been distributed to 1,267 doctors who had prescribed the drug to 19,822 patients, including 3,760 women of childbearing age. This evidence raised the question whether the "professional judgment" of the medical research community could still be trusted. The net legislative result of this outcry in the United States was the passage of the Kefauver-Harris Bill, known as the Drug Amendments of 1962 and signed by President John F. Kennedy on October 10, 1962.[26]

This law fundamentally altered the character of research both for the drug industry and for academic medicine. It required (for the first time) that there be proof of efficacy for drugs. Also, it required that the labeling of drugs fully disclose contraindications, precautions, and harmful side effects. Third, it strengthened the requirements for new drug application and suspended the procedure of automatic approval. Finally, the law required that the FDA impose comprehensive regulations on the clinical testing of new drugs. The result of this law was nothing less than the transformation of the FDA into the final arbiter of what constituted successful achievement in the realm of medical therapeutics.[27]

The tool that the FDA turned to in adjudicating the issue of therapeutic efficacy was the clinical trial. As the law stated:

> "Substantial evidence" [of drug efficacy] means evidence consisting of adequate and well-controlled investigations, including clinical investigations, by experts qualified by scientific training and experience to evaluate the effectiveness of the drug involved, on the basis of which it could fairly and responsibly be concluded by such experts that the drug will have the effect it purports or is represented to have under the conditions of use prescribed, recommended, or suggested in the labeling or proposed labeling thereof.[28]

With the passage of the 1962 law, the FDA set up specific guidelines for implementing the law. Among them was the requirement that the FDA approve any changes in the design of a clinical trial to test a new therapy. The conductor of the trial was required to submit a "reasonable protocol" developed on the basis of earlier phases of the investigation. This plan had to contain data concerning the specific nature of the investigation. It also had to provide for the utilization of more than one independent investigator who had to keep adequate records of the results of the study.[29]

With the advent of the clinical trial as a standard procedure that was subject to government regulation, "medical decision making" had emerged as an issue of public policy in most of the major Western industrial democracies. Although the drug regulatory agencies initially relied on the professional clinical judgment of the physicians on their staff, the concerns of the statistician took on increasing importance in assessing the results of the clinical trial. By the late 1960s the double-blind methodology had become mandatory for FDA approval in the United States, and the procedure had become standard in most of the other Western industrial democracies as well by the late 1970s.[30]

The expertise of the statistician was considered important because of the heterogeneity of modern pluralistic societies, in which results expressed numerically had greater authority whether these involved medical therapies, the outcome of elections, or risk analysis. They were seen as rising above the individual differences and possessing "objectivity" and "truth." The emergence of the clinical trial can thus be seen as a special case of a more general trend—the belief that "numbers rule the world."[31]

With the triumph of the clinical trial as a standard procedure, this story has reached its historical denouement. However, the procedure has not been immune from controversy. In the Conclusion I assess how contemporary debates over the proper design of clinical trials can be further illuminated by comparison with the prehistory of the procedure documented in this study, and I discuss these debates in the wider context of our "postmodern" culture.

CONCLUSION

THE GOVERNMENT REGULATIONS passed by most of the major Western industrial democracies within the past twenty to thirty years have had a profound effect on the prestige of the clinical trial. The magnitude of the impact of the clinical trial as an established and standard procedure is evident in the praise of Hill's work by one of his present-day admirers (he was knighted in 1961):

> Few innovations have made such an impact on medicine as the controlled clinical trial that was designed by Sir Austin Bradford Hill for the Medical Research Council's Streptomycin in Tuberculosis Trials Committee in 1946. Thirty-five years later the structure, conditions of conduct, and analysis of the currently standard trials are, for the most part, the same. Its durability is a monument to Sir Austin's scientific perception, common sense, and *concern for the welfare of the individual patient.*[1] [my emphasis]

In light of the criticisms of Louis that his "numerical method" denied the individuality of the patient, such encomiums to Hill clearly bespeak a sea change in medical and public opinion regarding the efficacy of a statistical methodology.

With the emergence of the clinical trial as a standard procedure, several pressing issues have been pushed to the forefront of public debate. How should the imperative to heal the sick patient be balanced against the need to expand the frontiers of scientific and medical knowledge? How much credence should be given to the "professional expertise" of the clinician or medical researcher and how much authority should be vested in the conclusions of the professional statistician? How does the advent of the clinical trial fundamentally alter the nature of the doctor-patient relationship? Or, more specifically, to what extent has society at large (through governmental regulation) rather than professional expertise become the primary vehicle for conferring "authority" on the decisions of the physician?[2] Finally, do medicine's "scientific" credentials derive from the use of laboratory techniques in such fields as physiology or bacteriology, or from more novel techniques of mathematical statistical inference as used in epidemiology?[3] As this study has made clear, however, these issues did not originate with the triumph of the clinical trial in the last thirty years; they were present every time that numerical comparison was used to determine the efficacy of therapy during the nineteenth and twentieth centuries.

Despite the intellectual similarity to earlier debates, the institutionalization of the clinical trial has also generated new medical, ethical, political, economic, and epistemological issues. One of the most pressing is the determination of groups of at-risk populations to be studied in designing a clinical trial. Should experimental subjects in the trials be composed largely of middle- and upper-class white males (as has been the case in the past), or should women and minorities also be included?[4] Viewed in this way, the design of the clinical trial can provide yet another opportunity to address that galloping troika of issues so beloved by the new social history—race, class, and gender.

Also, the thorny question of when to terminate a clinical trial can often be seen as a conflict between the epistemic value of doing "good science" and the ethical value of administering a potentially useful treatment if the preliminary results look promising. The clearest exposition of this dichotomy in literature appeared in Sinclair Lewis's 1925 novel *Arrowsmith*, in which the main character administered a new and experimental drug to the epidemic-ridden population (including the control group from which treatment had previously been withheld) when his wife died from the disease. Similar issues are being played out in the clinical trials that test new AIDS-related therapies. The *New York Times* raised the ethical problem of AIDS patients lying about previous treatment so that they can get admitted to trials in which new therapies are being administered. As one article asked rhetorically, "Can desperately ill people be expected to feel a moral obligation to the demanding rigors of the scientific method?"[5]

The clinical trial can also be seen as pitting economic and political interests against each other. Pharmaceutical companies have a vested economic interest in having clinical trials performed to see if the French-produced abortion pill RU-486 is truly efficacious. However, clinical trials to test this pill have only recently been approved in the United States because of the political firestorm surrounding anything involving abortion here.[6] In short the study of the design and implementation of the clinical trial provides a unique opportunity for engaging some of the most hotly contested issues of our day.

Underlying both the contemporary issues outlined above and the historical issues discussed in the preceding chapters has been a recurring question: To what extent does the use of numerical comparison confer authority or "objectivity" on therapeutic choices? Viewed in this way, the debates surrounding the rise of the clinical trial engage a much broader issue currently being discussed within the academy; the attempt to show the socially and culturally "negotiated" character of the concept of "objectivity" (as well as the "negotiated" character of therapeutic efficacy). For many commentators the refutation of the idea that there is a unique, culturally transcendent way of understanding the world is one hallmark of the transition from the "modern" to the "postmodern" worldview.[7]

In his introduction to a recent collection of essays devoted to this topic of "rethinking objectivity," Allan Megill has outlined four senses of the term *objectivity* currently in use within this multidisciplinary-academic debate: absolute, disciplinary, dialectical, and procedural. The absolute notion of objectivity held that there was one unique way of "representing things as they really are"; all inquirers of goodwill are destined to converge on this one correct way of seeing reality. The disciplinary notion of objectivity is tied up with Thomas Kuhn's famous notion of a paradigm; it refers to the claim by practitioners of a particular discipline to have authoritative jurisdiction over its area of competence. The particular groundings for disciplinary objectivity vary from discipline to discipline and they change over time as well within a discipline. Dialectical objectivity recognizes that there is a role to play for subjective judgment; human inquirers constitute objects as objects through their subjective encounter with the world. Finally, procedural objectivity focuses solely on the impersonality of procedure or a set of rules; it emphasizes that such rules remove the taint of subjective human judgment.[8] In my account the debates over the use of numerical comparison within medicine can be seen as primarily a debate between disciplinary (medicine) and procedural (statistics) notions of objectivity; nevertheless, I argue that all four senses of objectivity have, in one way or another, been touched on in this narrative.

The role of disciplinary constraints in inhibiting the introduction of quantitative methods within medical therapeutics can be seen with great clarity when one reviews the arguments of Risueño d'Amador, Vierordt, and Wright. Even though Risueño d'Amador was a clinician whereas the other two were laboratory-based in orientation, all three rejected numerical methods in favor of some form of "tacit knowledge."[9] This rejection never derived from disagreement with specific conclusions (although often that was indeed the case). Rather, they made the more general claim that, for any clinician and/or researcher who had acquired sufficient professional skill through experience, the use of statistical or formal methods was unneccesary. Risueño d'Amador showed no interest in the technical features of the "calculus of probabilities" and was more concerned with what such methods would do to undermine the role of the doctor as humanitarian healer in the physician-patient relationship. Similarly, Vierordt showed no aversion to using quantitative reasoning in his physiological research and even recognized that there was a formal mathematical logic to Radicke's test; however, he maintained that such formal criteria could be overruled by the "logic of facts" that the experienced physiological researcher eventually learned to perceive. Finally, even Wright did not criticize Pearson's or Greenwood's technical mathematical competence (he called their mathematical manipulations both "unerring" and "beyond my intellectual ken"). Rather, he maintained that the "mathematical error" could safely be ignored as inconsequential after the so-called functional error of the experimenter had been sufficiently reduced. The fact that such frequent rivals—a clinician, a research physiologist, and a bacteriologist[10]—could unite in opposition to a particular form of quantitative reasoning speaks volumes about the strong power of discipline-induced notions of objectivity within medical culture.

The supporters of numerical comparison were likewise not just arguing against specific therapeutic innovations such as bloodletting, exposure to sea air, or vaccine therapy; they were offering an alternative vision of objectivity based on the procedural criteria that numbers removed the taint of subjective human judgment. Louis contended that the difference between expressions such as "more or less," and "rarely or frequently," and numbers is "the difference of truth and error; of a thing clear and truly scientific on the one hand, and of something vague and worthless on the other."[11] Radicke affirmed that "all that it is necessary to know, in laying down a rule for the comparison of two series of observations is— what we may learn from the series themselves, viz., whether fluctuations exist in those series, and if so, how great those fluctuations are. An acquaintance with their causes is quite unimportant."[12] Likewise, Greenwood maintained:

> The biometrician, *qua* biometrician, *discloses the data upon which his conclusions rest, the methods by which he has analysed them, and the conclusions which, in his opinion, may be drawn* [italics in original]. He does not appeal to the experience of statistical experts, nor found

conclusions upon a (practical) consensus of (qualified) opinion. It is open to any reader to study his paper or not to study it, to decipher the hieroglyphics or not to decipher them.[13]

Finally, Hill, as the father of the modern clinical trial, could declare that "it [numerical comparison] demands to the greatest possible extent objective measurements of the results and the use of subjective assessments only under a strict and efficient control which will ensure an absence of bias."[14] In contrast to the opponents of numerical comparison who emphasized a kind of private or disciplinary-based form of knowledge, supporters always emphasized the quintessentially "public" character of their methods; their use of number permitted their results to be submitted to scrutiny.[15]

Given that the crux of the three debates centered on the contrast between "private" and "public" visions of objectivity, it is understandable that the vision of objectivity that triumphed on each occasion depended largely on which side controlled the terms of the debate. Clearly, the clinical physicians and medical researchers had the upper hand until the post-World War II period; the proponents of probabilistic and statistical considerations were always "outsiders" in some way to the mainstream of the medical profession. Gavarret had what could only be characterized as an anomalous education in the highly stratified world of early-nineteenth-century French higher education—he had both a degree from the Ecole Polytechnique and a medical degree. Similarly, Radicke was a meteorologist and astronomer by training, not a physician. Even though Greenwood had received a medical degree and as a student had worked under the physiologist Leonard Hill, he had the unique experience of being one of the first individuals to study under Karl Pearson at his biometrical laboratory.

Through the work of Hill and his associates in the late 1940s and throughout the 1950s, statistical methods slowly gained a foothold in the world of medical therapeutics. Ultimately, however, the clinical trial gained legitimacy through the realization by society at large that the decisions of the medical profession had to be regulated. Only when the issue of "medical decision-making" moved from the cloistered confines of professional medical expertise into the arena of open political debate (thrust there by highly potent, industrially produced drugs) could the advocates of statistical or procedural objectivity gain the upper hand. In this respect the triumph of the clinical trial nicely reflects the intimate connection between procedural objectivity and democratic political culture.[16] As long as the medical elite indisputably controlled the terms of the debate, statistical objectivity could be marginalized. However, once the determination of therapy had to be justified before the tribunal of an often-distrustful public opinion, the appeal to impersonal rules became a very effective tool in the adjudication of medical uncertainty.

The debate over formal rules versus tacit knowledge in science has thus proved to be far more than an "academic" dispute for professional historians, philosophers, and sociologists of science; its resolution can have profound consequences for public policy and will affect anyone who receives medication from a physician (and that would include large portions of humanity). With the triumph of the clinical trial, the clinician and the medical researcher neither altered nor silenced their criticisms; however, they were now arguing from a defensive posture. In Kuhnian terms the transformation constituted nothing less than a revolution in the dominant paradigm of what constituted "objectivity."

The triumph of the clinical trial also highlights the changing nature of the doctor-patient relationship over time.[17] In the foregoing debates between procedural and disciplinary notions of objectivity, the issues have been viewed entirely from the physician/medical researcher's point of view, turning less on how the physician/researcher viewed the patient and more on what sort of criteria should guide the action of the medical professional administering therapy; the patient was not taken seriously. However, the view of the patient has also undergone profound transformation during the course of the nineteenth and twentieth centuries. Just as the proper role for the medical professional could be seen as a conflict between disciplinary and procedural notions of objectivity, so likewise could the proper role for the patient be seen as a conflict between dialectical and absolutist notions of objectivity—the other two senses of objectivity outlined by Allan Megill.

At the time of the debates of the Paris Academy of Medicine, all the relevant participants viewed the patient in dialectical terms. Risueño d'Amador rejected the numerical method because he believed that it was the duty of the physician to recognize the idiosyncrasies of the individual patient and administer the therapy uniquely suited to that patient. In terms of Megill's formulation of dialectical objectivity, Risueño d'Amador recognized that his subjective impressions as a physician and the subjective impressions of the patient were both equally valid; the "objective" therapy emerged through the dialectical encounter between physician and patient. Louis likewise argued that the physician should take into account individuating factors such as the age, sex, or environment of the patient even when tabulating numerical results. Although Louis's focus remained the group rather than the individual, he did not deny (as some of his critics charged) that each patient was individual and unique.

During the course of the nineteenth century, this dialectical view of the patient was gradually replaced by an absolutist one for some physiological researchers and clinicians. In this study the views of Bernard and Wunderlich have been cited as representative examples. For Bernard, one of the reasons for rejecting the numerical method of Louis was that the laws of medicine had to "be based only on certainty, on *absolute* [my emphasis] determinism, not on probability."[18] Likewise Wunderlich could praise the thermometer because it produced observations that are "incontestable and indubitable, which are independent of the opinion or the amount of practice or the sagacity of the observer—in a word, materials which are physically accurate."[19] This view that precision measuring instruments rendered medical diagnosis "objective" (in the absolutist sense) was reflected concretely in medical practice; taking the temperature of patients was a task often delegated to nurses or other individuals who did not possess as authoritative status as physicians.[20] The data produced by the measuring instruments were seen as so unambiguous that no professional or discipline-induced criteria were needed to record the results. Disciplinary-based authority entered the picture only after the results had been recorded, when it was time to determine the diagnosis and administer therapy. The story of the thermometer proved to be a harbinger of things to come: the rise of "absolute objectivity" in "modern" medical diagnosis can be seen in the growing prestige given to results generated by a whole host of new (and often high-technology) precision measuring instruments during the present century.

As might have been expected, such "modernist" objectivism eventually begat a "postmodern" critique, particularly following the emergence of medical ethics as a new professional discipline. Now there are a multiplicity of academic voices calling for the abandonment of the typical "monologic" medical history that is "controlled by a scientific ideology that focuses on the biochemical aspects of a disease and its treatment to the exclusion of the human being whose body harbors the disease so reified."[21] Instead, the argument is adduced that one should make the clinical encounter and the process of diagnosis "dialogic," with the healer and patient interactively creating an understanding of the experience of illness.[22] With the reemergence of the view that medical diagnosis should take into account the subjective impressions of the individual patient, we have come full circle: François Double and Risueño d'Amador proved to be "postmodernists" *avant la lettre*!

As the preceding analysis has indicated, most of the debates discussed in this study can be seen as reflecting the contested nature of "objectivity." This insight emerges only from historical hindsight; for the actual historical actors, the issue more often turned on the contested nature of "science." At the time of the debate of the Paris Academy of Medicine, Louis saw medicine as an essentially empirical science composed of direct observation and comparison. Gavarret, by contrast, saw the "science" of medicine as an extension of the mathematical theory of probabilities. Their contrasting visions had contrasting fates: whereas Gavarret was forgotten because his method proved too technical (Bruno Latour would argue that he failed to build a sufficiently stable network), Louis was eventually forgotten because his numerical method became so "standard" in establishing therapeutic efficacy that its association with "scientific" reasoning was eventually lost. The way was open for Bernard to argue in favor of an alternative vision of the science of medicine based not on mere empirical fact gathering, but rather on the methods of laboratory experimentation. Finally, Greenwood could claim that the methods of the laboratory as championed by Bernard were no longer sufficient to guarantee medicine's scientific status; the methods of statistical correlation must also be deployed.

This issue of the multiplicity of meanings that could be attributed to the phrase "science of medicine" was further publicized by the governmental regulations institutionalizing the double-blind clinical trial as the standard method for determining therapeutic efficacy. The medical profession would now have to use statistics (a tool of the social sciences) just as much as it had formerly used the laboratory (a tool of the natural sciences). This transformation marked a major shift differentiating nineteenth- from twentieth-century medical thinking in the realm of medical therapeutics.[23] In the nineteenth century the relationship between the statistician and the laboratory worker had been one of antagonism and non-communication, as epitomized in this study by the views of Louis and Bernard, respectively. In the twentieth century, by contrast, the relationship between the statistician and the laboratory worker could be characterized as one of growing, although hesitant, cooperation. Greenwood and Pearl emphasized how experimental epidemiology required that the researcher know both laboratory methods and statistical techniques in order to chart the effects of factors such as age and sex on the spread of a disease through a population. Sir Austin Bradford Hill argued repeatedly that the modern clinical trial should be a collaborative effort between the clinician and the statistician from start to finish. This would ensure both that the medical facts were correctly observed and that the correct statistical inferences were drawn. The rise of the clinical trial can thus be seen as yet another contribution to a growing body of scholarly literature in the history of medicine which has emphasized how alternative visions of scientific reasoning have competed in varying social and medical contexts in the attempt to "frame disease."[24]

With the current proliferation of medical specialties and subspecialties, the issue of what constitutes "objectivity" or "science" within medicine has never been more hotly contested. Examples could be multiplied ad nauseam beyond those already cited in this conclusion (the whole issue of whether or not computer-generated artificial intelligence constitutes "objective" medical knowledge leaps to the mind[25]). Despite these very real differences, the players in the Academy of Medicine debate of 1837 proved to be remarkably prescient, as the historical record chronicled in this study has indicated. With medical issues currently in the forefront of public debate and the number of professional "players" in these debates growing almost exponentially (or so it often seems), future historians of our contemporary medical debates over objectivity and science will have to provide much more "thick description"[26] than was necessary to account for the comparatively small number of players who figured prominently in this study. Nevertheless (if I may dare to prophesy), they will find that some of the broader issues raised by the use of quantification in a medical context are remarkably similar to those broached initially at the Parisian Academy of Medicine. As the French proverb has it, "Plus ça change, plus c'est la même chose."

NOTES

INTRODUCTION

1. Henk J.H.W. Bodewitz, Henk Buurma, and Gerard H. de Vries, "Regulatory Science and the Social Management of Trust in Medicine," in Wiebe E. Bijker, Thomas P. Hughes, and Trevor Pinch, eds., *The Social Construction of Technological Systems* (Cambridge: The MIT Press, 1987), pp. 243–259.

2. The role of quantitative objectivity as an instance of the rhetoric of public political discourse is addressed by Theodore M. Porter, "Objectivity and Authority: How French Engineers Reduced Public Utility to Numbers," *Poetics Today*, 12 (1991): 248–52.

3. Charles E. Rosenberg has emphasized that the use of a more comparative perspective is a defining feature of more recent historical scholarship in the 1987 afterword to his *The Cholera Years* (Chicago: University of Chicago Press, 1987 [1962]); see esp. pp. 239–42.

4. See Michel Foucault, *The Birth of the Clinic: An Archeology of Medical Perception*, trans. A. M. Sheridan Smith (New York: Vintage Books, 1975 [1963]); on related themes (and from a less overtly "philosophical" perspective) see Erwin H. Ackerknecht, *Medicine at the Paris Hospital, 1794–1848* (Baltimore: Johns Hopkins University Press, 1967).

5. Although not applied to a specifically medical context, the shift in concern within science studies from a focus on what scientists say to what they do has characterized the recent collection of essays in Andrew Pickering, ed., *Science as Practice and Culture* (Chicago: University of Chicago Press, 1992).

6. The idea that scientific reasoning was often based on some form of private or tacit knowledge was first developed by Michael Polanyi, *Personal Knowledge* (Chicago: University of Chicago Press, 1957).

7. Examples of the scholarly monographs produced that deal with these issues include Ian Hacking, *The Emergence of Probability* (Cambridge: Cambridge University Press, 1975); Barbara J. Shapiro, *Probability and Certainty in Seventeenth-Century England* (Princeton: Princeton University Press, 1983); Theodore M. Porter, *The Rise of Statistical Thinking, 1820–1900* (Princeton: Princeton University Press, 1986); Stephen M. Stigler, *The History of Statistics: The Measurement of Uncertainty before 1900* (Cambridge: The Belknap Press of Harvard University Press, 1986); Lorraine J. Daston, *Classical Probability in the Enlightenment* (Princeton: Princeton University Press, 1988); Gerd Gigerenzer et al., *The Empire of Chance* (Cambridge: Cambridge University Press, 1989); Ian Hacking, *The Taming of Chance* (Cambridge: Cambridge University Press, 1990); Lorenz Krüger et al., eds., *The Probabilistic Revolution*, 2 vols. (Cambridge: MIT Press, 1987).

8. Hacking, *The Taming of Chance* pp. 180–88.

9. See Allan Megill, ed., "Rethinking Objectivity," *Annals of Scholarship* 8 and 9 (1991–1992): 301–477; 1–153.

CHAPTER ONE

PROBABLE KNOWLEDGE IN THE PARISIAN SCIENTIFIC
AND MEDICAL COMMUNITIES DURING THE
FRENCH REVOLUTION

1. See Charles C. Gillispie, *Science and Polity in France at the End of the Old Regime* (Princeton: Princeton University Press, 1980).

2. On the career of Vicq d'Azyr, see ibid., pp. 29–33, 194–203.

3. The relationship between constitutional reform and Condorcet's interest in probability is discussed in ibid., pp. 33–50.

4. Lorraine J. Daston, *Classical Probability in the Enlightenment* (Princeton: Princeton University Press, 1988), pp. 14–48.

5. Quoted in Keith Michael Baker, *Condorcet: From Natural Philoso phy to Social Mathematics* (Chicago: University of Chicago Press, 1975), p. 156.

6. See Condorcet, "Reception Speech at the French Academy (1782)," in Keith Michael Baker, ed., *Condorcet: Selected Writings* (Indianapolis: Bobbs-Merrill, 1976), p. 6.

7. Pierre-Simon Laplace, *A Philosophical Essay on Probabilities*, 6th ed., trans. Frederick Wilson Truscott and Frederick Lincoln Emory (New York: Dover, 1951), p. 2.

8. Daston, *CLassical Probability*, pp. 188–89.

9. Laplace, *Philosophical Essay*, p. 196.

10. Ibid., pp. 105–6.

11. Ibid., p. 144.

12. S. F. Lacroix, *Traité élémentaire du calcul des probabilités* (Paris: Bachelier, 1833), p. 222.

13. Martin S. Staum, *Cabanis: Enlightenment and Medical Philosophy in the French Revolution* (Princeton: Princeton University Press, 1980), 78.

14. Quoted in Andrea A. Rusnock, "On the Quantification of Things Human: Medicine and Political Arithmetic in Enlightenment England and France" (Ph.D. Dissertation, Princeton University, 1990), p. 111.

15. Pierre-Jean-Georges Cabanis, "De degré de certitude de la médecine," in *Oeuvres philosophiques de Cabanis*, vol. 1 (Paris: Presses Universitaires de France, 1956), p. 91.

16. Pierre-Jean-Georges Cabanis, "Rapports du physique et du moral de l'homme," in *Oeuvres philosophiques de Cabanis*, vol. 1, p. 110.

17. For a discussion of this issue in an American context, see John Harley Warner, *The Therapeutic Perspective: Medical Practice, Knowledge, and Identity in America* (Cambridge: Harvard University Press, 1986), pp. 1–7 and *passim*.

18. Jan Goldstein, *Console and Classify: The French Psychiatric Profession in the Nineteenth Century* (Cambridge: Cambridge University Press, 1987), pp. 67–69, 89.

19. Ibid., p. 102.

20. Philippe Pinel, "Resultats d'observations et construction des tables pour servir à determiner le degré de probabilité de la guérison des aliénés," *Institut de France, Mémoires classe des sciences mathématique et physique* 8 (1807): 199.

21. The views of Cabanis and others are discussed in Terence D. Murphy, "Medical Knowledge and Statistical Methods in Early-Nineteenth-Century France," *Medical History* 25 (1981): 301–19.

22. Philippe Pinel, *Traité medico-philosophique sur l'aliénation mentale,* 2nd ed. (Paris, 1809), pp. 402–3.

23. See Robert Fox and George Weisz, eds., *The Organization of Science and Technology in France, 1808–1914* (Cambridge: Cambridge University Press, 1980); Toby Gelfand, *Professionalizing Modern Medicine: Paris Surgeons and Medical Science and Institutions in the Eighteenth Century* (Westport, Conn.: Greenwood Press, 1980).

CHAPTER TWO
LOUIS'S NUMERICAL METHOD IN EARLY-NINETEENTH-
CENTURY PARISIAN MEDICINE: THE
RHETORIC OF QUANTIFICATION

1. Erwin H. Ackerknecht, *Medicine at the Paris Hospital, 1791–1848* (Baltimore: The Johns Hopkins University Press, 1967), p. 15.

2. See Russell C. Maulitz, *Morbid Appearances: The Anatomy of Pathology in the Early Nineteenth Century* (Cambridge: Cambridge University Press, 1987), esp. Part 1 on Paris, pp. 9–105.

3. Michel Foucault, *The Birth of the Clinic: An Archeology of Medical Perception,* trans. A. M. Sheridan Smith (New York: Vintage, 1973 [1963]), p. 144.

4. For a survey of the traditional historiography, see Ulrich Tröhler, "Quantification in British Medicine and Surgery 1750–1830, with Special Reference to Its Introduction into Therapeutics" (Ph.D. dissertation, University of London, 1978), pp. 15–17.

5. See Tröhler, "Quantification in British Medicine"; Andrea A. Rusnock, "The Quantification of Things Human: Medicine and Political Arithmetic in Enlightenment England and France," (Ph.D. dissertation, Princeton University, 1990).

6. Studies that engage the issue of the multiplicity of meanings of the term *science* within medicine have been surveyed most recently in John Harley Warner, "History of Science and History of Medicine," in *Conference on Critical Problems and Research Frontiers in History of Science and History of Technology* (Madison, Wisconsin: October 30–November 3, 1991), esp. pp. 397–402.

7. John E. Lesch makes the point that Magendie effectively divided his time between the concerns of the clinical physician and those of the research physiologist, in his *Science and Medicine in France: The Emergence of Experimental Physiology, 1790–1855* (Cambridge: Harvard University

Press, 1984), p. 167; Terence D. Murphy emphasizes that one of the reasons those advancing the cause of internal medicine advocated medicine's scientific vocation was that they resented the increasing state regulation of medical practice in "The French Medical Profession's Perception of Its Social Function between 1776 and 1830," *Medical History* 23 (1979): 259–78.

8. This view that Louis's approach could be seen as a continuation of the Enlightenment ideology has been emphasized in Terence D. Murphy, "Medical Knowledge and Statistical Methods in Early-Nineteenth-Century France," *Medical History* 25 (1981): 301–19.

9. Ivo Schneider, "Laplace and Thereafter: The Status of Probability Calculus in the Nineteenth Century," in Lorenz Krüger, Lorraine J. Daston, Michael Heidelberger, et al., eds., *The Probabilistic Revolution*, vol. 1: *Ideas in History* (Cambridge: The MIT Press, 1987), p. 204.

10. Pierre Simon Laplace, *A Philosophical Essay on Probabilities*, 6th ed., trans. Frederick Wilson Truscott and Frederick Lincoln Emory (New York: Dover, 1952), p. 196.

11. John H. Talbott, *A Biographical History of Medicine* (New York: Grune and Stratton, 1970), pp. 497–99; James H. Cassedy, *American Medicine and Statistical Thinking, 1800–1860* (Cambridge: Harvard University Press, 1984), pp. 60–61; William Coleman, *Death Is a Social Disease: Public Health and Political Economy in Early Industrial France* (Madison: University of Wisconsin Press, 1982), pp. 132–33.

12. Pierre-Charles-Alexandre-Louis, *Pathological Researches on Phthisis*, trans. Charles Cowan (Boston: Hilliard, Gray, 1836), p. lxvii.

13. Pierre-Charles-Alexandre-Louis, *Anatomical, Pathological and Therapeutic Researches upon the Disease Known under the Name of Gastro-Enterite Putrid, Adynamic, Ataxic, or Typhoid Fever, etc., Compared with the Most Common Acute Diseases*, vol. 1, trans. Henry I. Bowditch (Boston: Hilliard, Gray, 1836), p. 228.

14. Ibid., vol. 2, trans. Henry I. Bowditch (Boston: Isaac R. Butts, 1836), p. 389.

15. Ibid., p. 390.

16. Ibid., pp. 398–99.

17. Ibid., p. 405.

18. Ibid., p. 412.

19. Pierre-Charles-Alexandre Louis, *Researches on the Effects of Bloodletting in Some Inflammatory Diseases, and on the Influence on Tartarized Antimony and Vesication in Pneumonitis*, trans. C. G. Putnam (Boston: Hilliard, Gray, 1836), pp. 48–49.

20. Ibid., p. 9.

21. Ibid., p. 19.

22. Erwin H. Ackerknecht, "Broussais, or A Forgotten Medical Revolution," *Bulletin of the History of Medicine* 27 (1953): 322–33; Michel Foucault has argued that with Broussais's 1816 text, "the historical and concrete a priori of the modern medical gaze was finally constituted," in *The Birth of the Clinic*, p. 192.

23. Louis, *Researches on the Effects of Bloodletting*, pp. 57–58.

24. Ibid., p. 68.

25. Ibid., p. 70.

26. Ibid., p. 60.

27. Ibid., p. 62.

28. Ibid., p. 59.

29. Bruno Danvin, "De la methode numérique et de ses avantages dans l'étude de la médecine," *Thèses de l'Ecole de Médecine* (Paris, 1831): 28.

30. For the eighteenth-century background to this work, see Tröhler, "Quantification in British Medicine," pp. 346–83.

31. Bernard-Pierre Lecuyer, "The Statistician's Role in Society: The Institutional Establishment of Statistics in France," *Minerva* 25 (1987): 52–53; Jean Civiale, *Traité de l'affection calculeuse, ou recherches sur la formation, les caractères physiques et chimiques, les causes, les signes, et les effets pathologiques de la pierre et de la gravelle suives d'un Essai de statistique sur cette maladie* (Paris: Crochard et Comp, 1838), pp. 548–50.

32. Civiale, *Traité*, pp. 601–3.

33. Reported in François-Joseph Double, "Recherches de statistique sur l'affection calculeuse," *Comtes rendus de l'Académie des Sciences* 1 (1835): 172.

34. *Dictionnaire de biographie française*, ed. R. D'Amat and R. Limouzin-Lamothe (Paris: Letouzey et Ané, 1967), s.v. "Double, François-Joseph."

35. François-Joseph Double, "Recherches de statistique sur l'affection calculeuse," *Comptes rendus de l'Académie des Sciences* 1 (1835): 173.

36. Ibid., p. 176.

37. Thomas H. Laqueur, "Bodies, Details, and the Humanitarian Narrative," in Lynn Hunt, ed., *The New Cultural History* (Berkeley: University of California Press, 1989), p. 182.

38. Claude Navier, "Statistique appliquée à la médecine," *Comptes rendus de l'Académie des Sciences* 1 (1835): 247–50.

39. François-Joseph Double, "Statistique appliquée à la médecine," *Comptes rendus de l'Académie des Sciences* 1 (1835): 281.

40. John E. Lesch, "The Paris Academy of Medicine and Experimental Science, 1820–1848," in Frederick Laurence Holmes and William Coleman, eds., *The Investigative Enterprise: Experimental Physiology in Nineteenth Century Medicine* (Berkeley: The University of California Press, 1988), p. 102.

41. *Bulletin de l'Académie Royale de Médecine* 1 (1836-37): 5–8.

42. For a discussion of Quetelet's early career, see Theodore M. Porter, *The Rise of Statistical Thinking, 1820–1900* (Princeton: Princeton University Press, 1985), pp. 40–55.

43. Lambert Adolphe Jacques Quetelet, *A Treatise On Man and the Development of His Faculties*, trans. R. Knox (New York: Burt Franklin, Research Source Works Series #247, 1962), p. vi.

44. Ibid., p. vii.

45. Ibid., p. x.

46. Ibid., pp. 70–72.

47. Ibid., p. 99.

48. See B. Bru, "Poisson, le calcul des probabilités et l'instruction publique," in Michel Metivier, Pierre Costabel, and Pierre Dugac, eds., *Siméon-Denis Poisson et la science de son temps* (Palaiseau: Ecole Polytechnique, 1981), pp. 51–94.

49. Siméon-Denis Poisson, *Recherches sur la probabilité des jugements en matière criminelle et en matière civile* (Paris: Bachelier, 1837), pp. 7–12.

50. Ibid., pp. v–vi.

51. Ibid., p. 209.

52. Ibid., p. 205.

53. F. Schiller, "The Statistician-Patient Relationship in Two Centuries: 'E Pluribus Unum,'" *International Congress of the History of Medicine* 24 (Budapest, 1974), Acta 1976, vol. 1, p. 295.

54. *Bulletin de l'Académie Royale de Médecine* 1 (1836-37): 603.

55. Risueño d'Amador, *Mémoire sur le calcul des probabilités appliqué à la médecine* (Paris, 1837), p. 11.

56. Ibid., p. 17.

57. Ibid., pp. 24–25.

58. Ibid., p. 37.

59. Ibid., p. 78.; on the broader context of the view of statistics among political economists, see Claude Menard, "Three Forms of Resistance to Statistics: Say, Cournot, Walras," *History of Political Economy* 12 (1980): 524–41.

60. Risueño d'Amador, *Mémoire sur le calcule*, pp. 56–57.

61. Ibid., p. 105.

62. *Dictionnaire universel des contemporains*, 4th ed., ed. G. Vapereau, suppl. Léon Garnier (Paris: Hachette, 1873), s.v. "Dubois [d'Amiens], Frederick."

63. *Bulletin de l'Académie Royale de Médecine* 1 (1836–37): 687.

64. Ibid., p. 692.

65. Ibid., p. 702.

66. For the details of this closely related controversy before the Academy of Sciences, see Lorraine Daston, *Classical Probability in the Enlightenment* (Princeton: Princeton University Press, 1988), pp. 359–69.

67. *Bulletin de l'Académie Royale de Médecine* 1 (1836-37): 704.

68. Ibid., p. 708.

69. Ibid., pp. 741–46.

70. Jean-Baptiste Bouillaud, *Essai sur la philosophie médicale* (Paris: Libraire des Sciences Médicales, 1836), p. x.

71. T.-C.-E. Auber, *Traité de philosophie médicale, ou exposition des verités générales et fondamentales de la médecine* (Paris: Germer Baillière, 1839), pp. 42–43.

72. L.-H. Petit, "Bulletin: Mort de M. le professeur Gavarret," *Union médicale: Journal des intérêts scientifiques et practiques moraux et professionnels du corps médical*, 3rd series, 50 (1890): 325.

73. Jules Gavarret, *Principes généraux de statistique médicale* (Paris: Libraires de la Faculté de Médecine de Paris, 1840), p. xiv.

74. Ibid., p. 42.

75. Ibid., pp. 120–21.

76. Ibid., p. 101.

77. Ibid., pp. 255–58.

78. Ibid., pp. 25–26.

79. Ibid., pp. 140–42.

80. Ibid., pp. 286–88.

81. Ibid., p. 93.

82. Ibid., pp. 289–91, 295–98.

83. Ibid., p. 222.

84. "Review of 'Principes généraux de statistique médicale, ou développement des règles qui doivent présider à son emploi,'" *British and Foreign Medical Review* 12 (July, 1841): 5.

85. Ibid., pp. 20–21.

86. Casimir Broussais, *De la statistique appliquée à la pathologie et à la thérapeutique* (Paris: J.-B. Baillière, Libraire de l'Académie Royale de Médecine, 1840), p. 8.

87. Ibid., pp. 101–2.

88. *Biographisches Lexikon der hervorragenden Ärtze*, 2nd ed., ed. W. Haberling and H. Vierordt (Berlin and Vienna: Urban and Schwarzenberg, 1932–35), s.v. "François-Louis-Isidore Valleix," Jacques Raigne-Delorme, "Necrologie-M. Valleix," *Archives de médecine*, 5th series, 6 (1855): 249.

89. F.-L.-I. Valleix, "De l'application de la statistique à la médecine: Examen critique de l'ouvrage de M. Gavarret" *Archives générales de médecine* 8 (1840): 10.

90. Ibid., p. 17.

91. Ibid., pp. 26–27.

92. Jules Gavarret, "Application de la statistique à la médecine: Réponse à l'examen critique auquel M. le docteur Valleix a soumis, dans le numéro de mai 1840 des *Archives générales de médecine*, l'ouvrage de M. Gavarret intitulé Principes généraux de statistique médicale ou développement des règles qui doivent présider à son emploi," *L'Expérience* 6 (1840): 1–12.

93. "Revue analytique et critique des Journaux de médecine français," *Revue médicale française et etrangère, Journal des progrès de la médecine Hippocratique* 2 (1840): 243–46.

94. On Gavarret's life, see Laborde, "Bulletin Hebdomadaire: Le Professeur Gavarret," *La Tribune médicale*, 2nd series, 22 (1890): 577–80; *Dictionnaire de biographie française*, s.v. "Gavarret, Jules."

CHAPTER THREE
NINETEENTH-CENTURY CRITICS OF GAVARRET'S
PROBABILISTIC APPROACH

1. The principal sources used in locating evidence cited in this chapter were the *Index Catalogue of the Library of the Surgeon General's Office*, 1st series (Washington, D.C., 1880–95) and Oskar B. Sheynin, "On the History

of Medical Statistics," *Archive for History of Exact Sciences* 26 (1982): 241–80.

2. Elisha Bartlett, *An Essay on the Philosophy of Medical Science* (Philadelphia: Lea and Blanchard, 1844), dedication.

3. Erwin H. Ackerknecht has characterized Bartlett's essay as "the only systematic formulation of the philosophical approach of the great Paris clinical school of Bayle, Laennec, Chomel, Louis" in "Elisha Bartlett and the Philosophy of the Paris Clinical School," *Bulletin of the History of Medicine* 24 (1950): 43.

4. Bartlett, *An Essay on the Philosophy of Medical Sciences*, p. 159.

5. Ibid., p. 179.

6. Ibid., p. 164.

7. *Dictionary of National Biography* (Sir Leslie Steven and Sir Sidney Lee, eds. (Oxford and New York: Oxford University Press, 1992 [1921–27]), s.v. "Guy, William Augustus," pp. 35–36; Lawrence Goldman, "Statistics and the Science of Society in Early Victorian Britain; An Intellectual Context for the General Register Office," *Social History of Medicine* 4 (1991): 423–24.

8. See Michael J. Cullen, *The Statistical Movement in Early Victorian Britain* (New York: Harvester, 1975).

9. Susan Faye Cannon, *Science in Culture: The Early Victorian Period* (New York: Science History Publications, 1978), pp. 240–44; Lawrence Goldman, "The Origins of British 'Social Science': Political Economy, Natural Science, and Statistics, 1830–1835," *Historical Journal* 26 (1983): 594–609.

10. Quoted in Victor L. Hilts, "*Aliis Exterendum*, or, the Origins of the Statistical Society of London," *Isis* 69 (1978): 37.

11. William A. Guy, "On the Original and Acquired Meaning of the Term 'Statistics,' and on the Proper Functions of a Statistical Society," *Journal of the Statistical Society* 28 (1865): 492.

12. William A. Guy, "On the Value of the Numerical Method as Applied to Science, but especially to Physiology and Medicine," *Journal of the Statistical Society of London* 2 (1839): 34.

13. Ibid., pp. 27, 45.

14. Ibid., pp. 42–43.

15. *The Cyclopedia of Anatomy and Physiology*, ed. Robert B. Todd (London: Longman et al., 1849–52), s.v. "Statistics, medical."

16. Ibid, p. 813.

17. William A. Guy, "The Numerical Method, and Its Application to the Science and Art of Medicine," *British Medical Journal* (1860): 469, 553.

18. See William Coleman, *Death Is a Social Disease: Public Health and Political Economy in Early Industrial France* (Madison: University of Wisconsin Press, 1982), pp. 136–37.

19. See Bernard-Pierre Lecuyer, "Probability in Vital and Social Statistics: Quetelet, Farr, and the Bertillons," in Lorenz Krüger, Michael Heidelberger, Lorraine J. Daston, et al., eds., *The Probabilistic Revolution*, vol. 1, *Ideas in History* (Cambridge: MIT Press, 1987), pp. 317–35.

20. *Dictionnaire encyclopédique des sciences médicales,* ed. A. Dechambre, L. Hahn, et al. (Paris: P. Asselin, G. Masson, 1883), s.v. "Moyenne."

21. Quoted in C. C. Heyde and E. Seneta, *I. J. Bienaymé: Statistical Theory Anticipated* (New York: Springer-Verlag, 1977), p. 7.

22. Ibid., pp. 5–10.

23. Irenée Jules Bienaymé, "Calcul des probabilités: Applications à la statistique médicale," *Procès verbaux de la Société Philomatique de Paris* (1840): 12–13.

24. I. J. Bienaymé, "Considérations à l'appui de la découverte de Laplace sur la loi de probabilité dans la methode des moindres carrés," *Comtes rendus des séances de l'Académie des Sciences* 37 (1853): 310.

25. Quoted in Ivo Schneider, "Laplace and Thereafter: The Status of Probability Calculus in the Nineteenth Century," in Lorenz Krüger, Michael Heidelberger, and Lorraine J. Daston, eds., *The Probabilistic Revolution,* vol. 1, pp. 200–201.

26. *Biographisches Lexikon der hervorragenden Ärtze,* 2nd ed., ed. W. Haberling and H. Vierordt (Berlin: Urban and Schwarzenberg, 1932–35), s.v. "Raige-Delorme, Jacques."

27. *Dictionnaire de médecine, ou repertoire général des sciences médicales considerées sous le rapport théorique et pratique,* 2nd ed., ed. N.-P. Adelon et al. (Paris: Bechet jne, 1782–1862), s.v. "Statistique, médicale."

28. Amédée Dechambre, "La Statistique médicale," *Gazette hebdomadaire de médecine et de chirurgie* 11 (1874): 195.

29. Ibid., p. 211.

30. *Dictionnaire encyclopedique des sciences médicales,* s.v. "Statistique, médicale—Application à la médecine."

31. These developments are surveyed in Frederick L. Holmes and William Coleman, eds., *The Investigative Enterprise: Experimental Physiology in Nineteenth-Century Medicine* (Berkeley: University of California Press, 1988), esp. Introduction, pp. 1–14.

32. Carl Wunderlich and A. Roser, "Einleitung: Über die Mängel der heutigen deutschen Medicin und über die Nothwendigkeit einer entschieden wissenschaftlichen Richtung in derselben," *Archiv für physiologische Heilkunde* 1 (1842): 3; I would like to acknowledge Kirsten Fisher and Gary A. Smith for assistance in the German translations in this chapter.

33. Ibid., p. 16.

34. *Biographisches Lexikon der Hervorragenden Ärtze,* s.v. "Wunderlich, Karl Reinhold August W."; "O. Heubner," "C. A. Wunderlich: Nekrolog," *Archiv der Heilkunde* 19 (1878): 289–320; Ackerknecht, "Broussais, or A Forgotten Medical Revolution," *Bulletin of the History of Medicine* 27 (1953): 324.

35. Carl August Wunderlich, "Das Verhältniss der physiologischen Medicin zur ärztlichen Praxis," *Archiv für physiologische Heilkunde* 4 (1845): 13.

36. G. Schweig, "Auseinandersetzung der statistischen Methode in besonderem Hinblick auf das medicinische Bedürfniss," *Archiv für physiologische Heilkunde* 13 (1854): 305.

37. Ibid., p. 322.

38. Ibid., pp. 325–26.

39. *Dictionary of Scientific Biography*, ed. Charles P. Gillispie (New York: Scribner, 1970–), s.v. "Von Pettenkofer, Max Josef."

40. Ludwig Seidel, "Ueber den numerischen Zusammenhang welcher zwischen der Häufigkeit der Typhus-Erkrankungen und dem Stande des Grundwassers während der letzten 9 Jahre in München hervorgetreten ist," *Zeitschrift für Biologie* 1 (1865): 226.

41. Ibid., p. 227.

42. See below, pp. 52–53.

43. Willers Jessen, "Zur analytischen Statistik," *Zeitschrift für Biologie* 3 (1867): 136.

44. *Biographisches Lexikon der Hervorragenden Ärtze*, s.v. "Oesterlen, Friedrich."

45. Friedrich Oesterlen, *Handbuch der medicinischen Statistik* (Tübingen: H. Laupp'schen, 1874), p. ix.

46. Ibid., p. 3.

47. Ibid., pp. 4, 25.

48. Ibid., pp. v, 1.

49. Ibid., p. 54.

50. C. Voit, "Adolph Fick (obituary)," in *Sitzungsberichte der mathematisch-physikalischen der k. b. Akademie der Wissenschaften zu München* (1902), pp. 277–87.

51. Adolf Fick, *Die Medicinische Physik* (Braunschweig: Friedrich Vieweg und Sohn, 1866), p. 430.

52. Ibid., p. 432.

53. Ibid., p. 441.

54. Ibid., p. 445.

55. Alexander Jokl, "Julius Hirschberg," *American Journal of Ophthalmology* 48 (1959): 330–31.

56. Julius Hirschberg, *Die mathematischen Grundlagen der Medicinischen Statistik, elementar Dargestellt* (Leipzig: Veit, 1874), pp. v–vi.

57. Ibid., p. ix.

58. Ibid., p. x.

59. Ibid., p. xi.

60. Ibid., p. xii.

61. Ibid., p. 48.

62. Ibid., p. 92.

63. Ibid., p. 94

64. "The Application of Mathematical Formulae to Medical Statistics," *British Medical Journal* (1875): 72–73.

65. Carl Liebermeister, "Ueber Wahrscheinlichkeitsrechnung in Anwendung auf therapeutische Statistik," *Volkmanns Sammlung klinischer Vorträge*, No. 110: Innere Medicin, No. 39 (1877): 937.

66. Ibid., p. 940

67. Friedrich Martius, "Autobiography," in L. R. Grote, ed., *Die Medizin der Gegenwart in Selbstdarstellungen* (Leipzig: Felix Meiner, 1923), pp. 107–10, 112–13, 138–40.

68. Friedrich Martius, "Die Principien der wissenschaftlichen Forschung in der Therapie," *Volkmanns Sammlung klinischer Vorträge* No. 139: Innere Medicin, No. 47 (1878): 1,172.

69. Ibid., p. 1,185.

70. Ibid., p. 1,188.

71. Friedrich Martius, "Die Numerische Methode (Statistik und Wahrscheinlichkeitsrechnung) mit besonderer Berücksichtigung ihrer Anwendung auf die Medicin," *Virchows Archiv für pathologische Anatomie und Physiologie und für klinische Medicin*, 83 (1881): 336.

72. Martius, "Autobiography," pp. 110–11.

73. Martius, "Die Numerische Methode," p. 337.

74. Ibid., p. 346.

75. Ibid., p. 366.

76. Ibid., p. 375.

77. Ibid., p. 376.

78. Ibid., p. 377.

79. Theodore M. Porter has viewed this divorce between statistical thinking and mathematical probability as characteristic of nineteenth-century thought more generally. Most of the standard methods of mathematical statistics have been developed since 1893. See *The Rise of Statistical Thinking, 1820–1900* (Princeton: Princeton University Press, 1986), esp. p. 315.

CHAPTER FOUR
THE LEGACY OF LOUIS AND THE RISE OF PHYSIOLOGY:
CONTRASTING VISIONS OF MEDICAL "OBJECTIVITY"

1. Louis's impact on the history of nineteenth-century American medicine has been the subject of considerable attention in the historical literature from the early twentieth century to the present day. See William Osler, "Influence of Louis on American Medicine," *Bulletin of the Johns Hopkins Hospital* 8 (1897): 161–67; Erwin H. Ackerknecht, "Elisha Bartlett and the Philosophy of the Paris Clinical School," *Bulletin of the History of Medicine* (1950): 43–60; John Harley Warner, *The Therapeutic Perspective: Medical Practice, Knowledge, and Identity in America 1820–1885* (Cambridge: Harvard University Press, 1986): on Louis, see esp. pp. 199–206.

2. This information was compiled by observing the list of members reported in each addition of the *Mémoires de la Société Médicale d'Observation*.

3. On Louis's retirement, see Major Greenwood, *The Medical Dictator and Other Biographical Studies* (London: Williams and Norgate, 1936), pp. 126–27.

4. Louis, "De l'examen des malades et de la recherche des faits généraux" *Mémoires de la Société Médicale d'Observation*, vol. 1 (Paris: Crochard, 1837), p. 3.

5. Ibid., p. 25.

6. Ibid., p. 27.

7. François-Louis-Isidore Valleix, "Avant-propos du deuxieme volume,"

Mémoires de la Société Médicale d'Observation de Paris, vol. 2 (Paris: Chez Fortin, Masson et Cie, 1844), p. xviii.

8. Ibid., pp. xix-xx.

9. Ibid., p. xxiii.

10. Valleix, "Recherches sur la frequence du pouls chez les enfants nouveau-nés et chez les enfants agés de sept mois à six ans," *Mémoires de la Société Médicale d'Observation de Paris*, vol. 2 (Paris: Chez Fortin, Masson et Cie, 1844), pp. 300–80.

11. On the relationship of Flint and his son (also a doctor) to the issue of the proper professional training of the physician, see John Harley Warner, "Ideals of Science and Their Discontents in Late-Nineteenth-Century American Medicine," *Isis* 82 (1991): 471–72.

12. Austin Flint, "Remarks on the Numerical System of Louis," *New York Journal of Medicine and Science* 4 (1841): 284.

13. William and Daniel Griffin, *Medical and Physiological Problems; being chiefly researches for Correct Principles of Treatment in Disputed Points of Medical Practice* (London: Sherwood, Gilbert and Piper, 1843), pp. 224–26.

14. Ibid., p. 233.

15. François-Joseph Double, "Introduction," *Traité de médecine-pratique de Jean-Pierre Frank* (Paris: J.-B. Baillière Libraire de l'Académie Royale de Médecine, 1842) p. xxxv.

16. Ibid., p. xxxvi.

17. Friedrich Oesterlen, *Medical Logic*, trans. G. Whitley (London: Sydenham Society, 1855 [1852]), p. 284.

18. William Pulteney Alison, "Notes on the Application of Statistics to Questions in Medical Science, Particularly as to the External Causes of Diseases," *Report of the BAAS* (1855): 155.

19. Ibid., p. 156.

20. *Biographisches Lexikon der Hervorragenden Ärtze*, s.v. "Trousseau, Armand T."; Louis, "Recherches sur la fièvre jaune de Gibraltar de 1828," *Mémoires de la Société Médicale d'Observation de Paris*, vol. 2 (Paris: Chez Fortin, Masson et Cie, 1844), pp. 298–99.

21. Armand Trousseau, "Introduction—What Is Clinical Medicine," in *Lectures on Clinical Medicine, Delivered at the Hôtel-Dieu*, Paris, trans. John Rose Cormack from 3rd revised edition of 1868, vol. 2 (London: The New Sydenham Society, 1869), p. 34.

22. David W. Cheever, "The Value and the Fallacy of Statistics in the Observation of Disease," *The Boston Medical and Surgical Journal* 63 (1861): 515.

23. Adolphe Quetelet, *Letters Addressed to H.R.H. The Grand Duke of Saxe Coburg and Gotha*, trans. Olinthus Gregory Downes (London: Charles and Edwin Layton, 1849), p. 228.

24. Ibid., p. 229.

25. Ibid., p. 232.

26. Ibid., p. 235.

27. John Herschel, "Letters Addressed to H.R.H. the Grand Duke of Saxe-Cobourg and Gotha on the Theory of Probabilities as Applied to the Moral and Political Sciences," *Edinburgh Review* 92 (1850): 11.

28. Ibid., p. 52.

29. Ibid., p. 54.

30. Joseph Lister, "Effects of the Antiseptic System of Treatment upon the Salubrity of a Surgical Hospital," *The Lancet* i (1870): 40; according to modern statistical theory, the difference in the average values is statistically significant according to the chi-squared test, on which see Stuart J. Pocock, *Clinical Trials: A Practical Approach* (New York: Wiley, 1983), p. 16.

31. Lister, "Further Evidence Regarding the Effects of the Antiseptic System of Treatment upon the Salubrity of a Surgical Hospital," *The Lancet* ii (1870): 288.

32. See Ian Hacking, *The Taming of Chance* (Cambridge: Cambridge University Press, 1990), pp. 160–69.

33. Auguste Comte, *Cours de philosophie positive*, 2nd ed., vol. 3 (Paris: J. B. Baillière, 1864), p. 292.

34. On the interactive relationship between Bernard and Parisian clinical medicine, see John E. Lesch, *Science and Medicine in France: The Emergence of Experimental Physiology, 1790–1855* (Cambridge: Harvard University Press, 1984).

35. For an analysis of Bernard as a discipline-builder, see William Coleman, "The Cognitive Basis of the Discipline: Claude Bernard on Physiology," *Isis* 76 (1985): 49–70.

36. Claude Bernard, *An Introduction to the Study of Experimental Medicine*, trans. Henry Copley Greene (New York: Dover, 1957), p. 146.

37. Ibid., pp. 147–48.

38. Ibid., p. 148.

39. Ibid., p. 62.

40. Claude Bernard, *Principes de médecine expérimentale* (Paris: Presses Universitaires de France, 1947), p. 50.

41. Bernard, *An Introduction to the Study of Experimental Medicine*, pp. 129–40; For an analysis of Bernard's views on statistics, see Joseph Schiller, "Claude Bernard et la statistique," *Archives internationales d'histoire des sciences* 16 (1963): 405–18.

42. Bernard, *An Introduction to the Study of Experimental Medicine*, p. 135.

43. Bernard, "De l'emploi des moyennes en physiologie expérimentale à propos de l'influence de l'effeuillage des betteraves sur la production," *Comptes rendus de l'Académie des Sciences* 81 (1875): 702–3.

44. Zeno G. Swijtink, "The Objectification of Observation: Measurement and Statistical Methods in the Nineteenth Century," in Lorenz Krüger, Lorraine J. Daston, Michael Heidelberger, et al., eds., *The Probabilistic Revolution*, vol. 1 (Cambridge: MIT Press, 1987), pp. 261–85.

45. On the rise of the view of health as defined by physiological norms (such as temperature) and the concomitant rise of instrumentation, see

John Harley Warner, *The Therapeutic Perspective: Medical Practice, Knowledge, and Identity in America, 1820–1885* (Cambridge: Harvard University Press, 1986), pp. 87–101; see also Robert G. Frank, Jr., "The Telltale Heart: Physiologicial Instruments, Graphic Methods, and Clinical Hopes, 1854–1914," in Frederick Laurence Holmes and William Coleman, eds., *The Investigative Enterprise: Experimental Physiology in Nineteenth-Century Medicine* (Berkeley: University of California Press, 1988), pp. 201–90.

46. Stanley Joel Reiser, *Medicine and the Reign of Technology* (Cambridge: Cambridge University Press, 1978), pp. 115–21.

47. Carl Auguste Wunderlich, *On the Temperature in Diseases: A Manual of Medical Thermometry*, trans. W. Bathurst Woodman (London: New Sydenham Society, 1871 [1868]), pp. 48–49.

48. Ibid., p. 51.

49. Ibid., p. 289.

50. Ibid.

51. Ibid.

52. John S. Billings, "Our Medical Literature," *Lancet* ii (1881): 269.

53. Ibid.

54. Gustav Radicke, "On the Importance and Value of Arithmetic Means," trans. Francis T. Bond (London: The New Sydenham Society, 1861), 185.

55. William Coleman, "Experimental Physiology and Statistical Inference: The Therapeutic Trial in Nineteenth-Century Germany," in Krüger, Daston, Heidelberger, et al., eds., *The Probabilistic Revolution*, vol. 2 (Cambridge: MIT Press, 1987), pp. 206–7.

56. Radicke, "On the Importance and Value of Arithmetic Means," p. 186.

57. Ibid., p. 189.

58. Ibid., p. 193.

59. Ibid., p. 195.

60. Ibid., p. 200.

61. Ibid., p. 218.

62. Ibid., pp. 220–21.

63. Ibid., pp. 222–23.

64. Ibid., p. 250.

65. Reiser, *Medicine and the Reign*, pp. 131–32; Karl Vierordt, "Counting Human Blood Corpuscles," in Logan Clendening, ed., *Source Book of Medical History* (New York: Dover, 1960 [1942]), pp. 581–85.

66. Vierordt, "Notes on Medical Statistics," trans. Francis T. Bond (London: The New Sydenham Society, 1861), p. 252.

67. Ibid., p. 255.

68. F. W. Beneke, "A Reply to Professor Radicke's Paper, 'On The Importance and Value of Arithmetic Means,'" trans. Francis T. Bond (London: New Sydenham Society, 1861), p. 258.

69. Ibid., pp. 259–60.

70. Ibid., p. 261.

71. Radicke, "On the Deduction of Physiological and Pharmaco-Dynami-

cal Probabilities from Co-ordinated Series of Observations," trans. Francis T. Bond (London: The New Sydenham Society, 1861), p. 270.

72. Ibid., p. 271.

73. Ibid., p. 272.

74. Ibid., p. 273.

75. William Coleman has attributed the physiologists' rejection of Radicke to broader philosophical currents within mid-nineteenth-century German science, such as the distrust of abstract theorizing following the failure of the program of Naturphilosophie and the commitment to physiological determinism in his article "Experimental Physiology and Statistical inference: The Therapeutic Trial in Nineteenth-Century Germany," in *The Probabilistic Revolution*, vol. 2 (Cambridge: MIT Press, 1987), pp. 201–26. Although I agree that determinism provided the general philosophical framework for justifying a laboratory-based orientation, I believe that the rejection of Radicke, in particular, stemmed more from a desire to resist the limitations that formal rules would introduce into research practice. Some of the advantages of studying science-as-practice rather than science-as-knowledge are surveyed in Andrew Pickering, ed., *Science as Practice and Culture* (Chicago: University of Chicago Press, 1992).

76. "The Application of Mathematical Formulae to Medical Statistics," *British Medical Journal* (1875): 72.

CHAPTER FIVE
THE BRITISH BIOMETRICAL SCHOOL AND BACTERIOLOGY:
THE CREATION OF MAJOR GREENWOOD
AS A MEDICAL STATISTICIAN

1. Historical accounts that explicitly treat the emergence of statistics from this perspective include Donald A. MacKenzie, *Statistics in Britain, 1865–1930: The Social Construction of Scientific Knowledge* (Edinburgh: Edinburgh University Press, 1981), esp. p. 94; Joseph Ben-David, *The Scientist's Role in Society: A Comparative Study* (Englewood Cliffs: Prentice Hall, 1971), pp. 147–68; Gerd Gigerenzer et al., *The Empire of Chance: How Probability Changed Science and Everyday Life* (Cambridge: Cambridge University Press, 1989), pp. 109–22.

2. Stephen M. Stigler, *The History of Statistics: The Measurement of Uncertainty before 1900* (Cambridge: The Belknap Press of Harvard University Press, 1986), pp. 266–67.

3. Quoted in Theodore M. Porter, *The Rise of Statistical Thinking, 1820–1900* (Princeton: Princeton University Press, 1986), pp. 134–35.

4. See Frank M. Turner, *Between Science and Religion: The Reaction to Scientific Naturalism in Late Victorian England* (New Haven: Yale University Press, 1974), pp. 8–37.

5. "Miscellanea," *Journal of the Statistical Society* 40 (1877): 471.

6. Ibid., p. 472.

7. Ibid., p. 475.

8. See Victor L. Hilts, "Statistics and Social Science," in Ronald Giere

and Richard Westfall, eds., *Foundations of the Scientific Method: The Nine-teenth Century* (Bloomington: Indiana University Press, 1973), pp. 206–33; Ruth S. Cowan, "Francis Galton's Statistical Ideas: The Influence of Eugen-ics," *Isis* 63 (1972): 509–28.

9. Ian Hacking, *The Taming of Chance* (Cambridge: Cambridge Univer-sity Press, 1990), pp. 181–88.

10. Stigler, *History of Statistics*, p. 266.

11. Karl Pearson in *Speeches Delivered at a Dinner Held in University Col-lege, London, in Honour of Professor Karl Pearson 23 April, 1934* (Printed Pri-vately at the University Press, Cambridge, 1934) in box 39 of the Karl Pear-son Papers, University College, London. Subsequently cited as KPP.

12. Karl Pearson, *The Grammar of Science*, 3rd edition (New York: Macmillan, 1911), p. 37.

13. Ibid., pp. 164–65

14. Karl Pearson to Francis Galton, December 13, 1900, box 293D, in Francis Galton Papers, University College, London. Subsequently cited as FGP.

15. Pearson to Galton, April 18, 1901, FGP, box 293E.

16. Pearson to Galton, April 30, 1901, FGP, box 293E.

17. Galton to Pearson, February 28, 1902, FGP, box 245/18E.

18. Galton to Pearson, April 23, 1901 and December 8, 1902, FGP, box 245/18E.

19. University College, Senate Minutes, October 1904, FGP, box 131/1.

20. MacKenzie, *Statistics in Britain*, pp. 101–4.

21. Pearson to Galton, February 7, 1909, FGP, box 293K.

22. Quoted in Daniel J. Kevles, *In the Name of Eugenics: Genetics and the Uses of Human Heredity* (Berkeley: University of California Press, 1985), p. 59.

23. Pearson to Galton, February 7, 1909, FGP, box 293K.

24. Pearson to Galton, April 6, 1909, FGP, box 293K.

25. Pearson to Galton, July 3, 1901, FGP, box 283E.

26. Galton to Pearson, July 4, 1901, FGP, box 245/18E.

27. Galton to Pearson, July 8, 1901, FGP, box 245/18E.

28. Pearson to Galton, January 11, 1898, FGP, box 293C.

29. Even though the biologists were skeptical, the contemporary physics community was more actively embracing a statistical approach to deal with problems in kinetic gas theory as developed in the second half of the nine-teenth century by Maxwell and Boltzmann. See Porter, *The Rise of Statisti-cal Thinking*, pp. 110–28, 193–227.

30. Pearson to Galton, March 5, 1903, FGP, box 293F.

31. Pearson to Galton, May 13, 1906, FGP, box 283G.

32. Galton to Pearson, May 14, 1906, FGP, box 245/18G; Pearson to Gal-ton, June 29, 1906, FGP, box 293G.

33. Galton to Pearson, April 23, 1901, FGP, box 245/18E.

34. Pearson to Galton, February 7, 1909, FGP, box 293K.

35. Pearson to Galton, February 10, 1909, FGP, box 293K.

36. Christopher Lawrence, "Incommunicable Knowledge: Science, Tech-nology and the Clinical Art in Britain, 1850–1914," *Journal of Contemporary*

History 20 (1985): 503–20; on related issues, see S.E.D. Shortt, "Physicians, Science, and Status: Issues in the Professionalization of Anglo-American Medicine in the Nineteenth Century," *Medical History* 27 (1983): 59.

37. On the interface between the laboratory and the clinic, see Russell C. Maulitz, "'Physician versus Bacteriologist': The Ideology of Science in Clinical Medicine," in Morris J. Vogel and Charles E. Rosenberg, eds., *The Therapeutic Revolution: Essays in the Social History of American Medicine* (Philadelphia: University of Pennsylvania Press, 1979), pp. 91–107.

38. Philip Henry Pye-Smith, "Medicine as a Science and Medicine as an Art," *The Lancet* ii(1900): 309.

39. Ibid., p. 310.

40. Sir Humphry Davy Rolleston, *The Right Honourable Sir Thomas Clifford Allbutt* (London: Macmillan, 1929), pp. 12, 34, and 56–58.

41. T. Clifford Allbutt, "An Introductory Address on Theory and Practice," *The Lancet* ii (1899): 926.

42. Ibid., p. 926.

43. Ibid., p. 928.

44. Sir Thomas Clifford Allbutt, ed., Introduction, *A System of Medicine by Many Writers*, vol. I (New York: Macmillan, 1901), p. xxx.

45. Ibid., p. xxxix.

46. On the nineteenth-century origins of the concept of a "normal state," see Hacking, *The Taming of Chance*, pp. 160–69.

47. See Zeno G. Swijtink, "The Objectification of Observation: Measurement and Statistical Methods in the Nineteenth Century," in Lorenz Krüger, Lorraine J. Daston, Michael Heidelberger, et al., eds., *The Probabilistic Revolution*, vol. 1 (Cambridge: MIT Press, 1987), pp. 261–85.

48. Leonard Colebrook, *Almroth Wright: Provocative Doctor and Thinker* (London: William Heinemann, 1954), pp. 11–18.

49. *The London Times*, October 27, 1904, p. 7; *Munk's Role. Lives of the Fellows of the Royal College of Physicians of London*, ed. R. R. Trail (Oxford: IRL Press and Oxford University Press, 1968), vol. 5., s.v. "Wright, Sir Almroth Edward."

50. A. E. Wright, *A Short Treatise on Anti-Typhoid Inoculation* (Westminster: Archibald Constable, 1904), p. 43.

51. Ibid., p. 47.

52. Ibid., pp. 48–50.

53. Karl Pearson to Lt.-Colonel R.J.S. Simpson, C.M.G., R.A.M.C., May 24, 1904, KPP, box 159/1.

54. Simpson to Pearson, May 26 and 30, 1904, KPP, box 159/1.

55. Modern statistical theory would also accept the possibility of negative values of r up to $r = -1$; however, Pearson did not discuss negative correlation in his *British Medical Journal* article.

56. Pearson, "Report of Certain Enteric Fever Inoculation Statistics," *British Medical Journal* (1904): 1,244.

57. Ibid.

58. "Antityphoid Inoculation," *British Medical Journal* (1904): 1,260.

59. Ibid., p. 1,261.

60. Wright, "Antityphoid Inoculation," *British Medical Journal* (1904): 1,344.

61. Ibid., pp. 1,344–45.

62. Pearson, "Antityphoid Inoculation," *British Medical Journal* (1904): 1,432.

63. Wright, "Antityphoid Inoculation," 1,490.

64. Pearson, "Antityphoid Inoculation," 1,667.

65. Wright, "Antityphoid Inoculation," 1,727.

66. Pearson, "Antityphoid Inoculation," 1,776.

67. Zachary Cope, *Almroth Wright: Founder of Modern Vaccine-Therapy* (London: Thomas Nelson, 1966), pp. 39–45.

68. Quoted in ibid., p. 47.

69. Wright, "On Some Points in Connection with Vaccine-Therapy and Therapeutic Immunisation Generally," *The Practitioner* 80 (1908): 605.

70. Quoted in Colebrook, *Almroth Wright: Provocative Doctor and Thinker*, p. 267.

71. Quoted in ibid., pp. 265–66.

72. "Research in Medicine," *The Lancet* ii (1905): 772.

73. "The Medicine of the Future," *The British Medical Journal* (1907): 333.

74. See Greenwood's autobiographical statement (1924) in Greenwood file, Raymond Pearl Papers, American Philosophical Society, Philadelphia, Pennsylvania, pp. 1–8. Subsequently cited as RPP.

75. Greenwood to Pearson, March 18, 1902, KPP, box 707.

76. Greenwood to Pearson, September 21, 1904; Pearson to Greenwood, September 22, 1904, KPP, boxes 707 and 915, respectively.

77. Greenwood to Pearson, September 25, 1904, KPP, box 707.

78. See Greenwood's autobiographical statement (1924) in RPP, p. 5; Lancelot Hogben, "Major Greenwood, 1880–1949," *Obituary Notices of Fellows of the Royal Society* (1950): 139–41.

79. Greenwood to Pearson, January 19, 1908, KPP, box 707.

80. Greenwood, "Statistical Considerations Relative to the Opsonic Index," *The Practitioner* 80 (1908): 643.

81. J.D.C. White and Major Greenwood, "A Biometric Study of Phagocytosis with Special Reference to the 'Opsonic Index,'" *Biometrika* 6 (1908-1909): 377.

82. Ibid., p. 378.

83. Ibid., p. 401; see also White and Greenwood, "A Biometric Study of Phagocytosis with Special Reference to the 'Opsonic Index,' Second Memoir. On the Distribution of the Means of Samples," *Biometrika* 7 (1909-1910): 505–30.

84. Greenwood, "A Statistical View of the Opsonic Index," *Proceedings of the Royal Society of Medicine* 2 (1909): 146.

85. Ibid., pp. 149–54; on the idea of using the mode rather than the mean, see Pearson to Greenwood, December 4, 1908, KPP, box 915.

86. Greenwood, "A Statistical View of the Opsonic Index," p. 155.

87. Ibid., pp. 157–58.

88. W. F. Harvey and Anderson McKendrick, "The Opsonic Index—A Medico-Statistical Enquiry," *Biometrika* 7 (1909-1910): 64.

89. Wright, "Vaccine Therapy: Its Administration, Value, and Limitations. An Address Introductory to a Discussion on the Subject," *Proceedings of the Royal Society of Medicine*, 3 (1910): 4. *Note: Documentation for notes 89–96 can be found under "Vaccine Therapy" in the Primary Sources section of the Bibliography.*

90. Ibid., p. 8.

91. Arthur Latham, "Discussion on 'Vaccine Therapy: Its Administration, Value, and Limitations,'" *Proceedings of the Royal Society of Medicine*, 3 (1910): 128.

92. T. J. Horder, "Discussion on 'Vaccine Therapy: Its Administration, Value, and Limitations,'" *Proceedings of the Royal Society of Medicine*, 3 (1910): 139.

93. William Bulloch, "Discussion on 'Vaccine Therapy: Its Administration, Value, and Limitations,'" *Proceedings of the Royal Society of Medicine*, 3 (1910): 79.

94. Ibid., pp. 79–80.

95. Wright, "Vaccine Therapy: Its Administration, Value, and Limitations," *Proceedings of the Royal Society of Medicine*, 3 (1910): 29.

96. Leonard Colebrook, "Vaccine Therapy: Its Administration, Value, and Limitations," *Proceedings of the Royal Society of Medicine*, 3 (1910): 44.

97. KPP, box 399.

98. Pearson to Greenwood, end of November, 1910, KPP, box 915.

99. Pearson, "The Opsonic Index—'Mathematical Error and Functional Error,'" *Biometrika* 8 (1911–1912): 204.

100. Ibid., pp. 221.

101. Greenwood to Pearson, November 15, 1912, KPP, box 707.

102. Greenwood to Pearson, January 19, 1908, KPP, box 707.

103. Greenwood to Ronald Ross, April 16, 1910, Ross papers, London School of Hygiene and Tropical Medicine, file 74, running number 74/023.

104. "Mathematics and Medicine," *British Medical Journal* (1911): 449.

105. Pearson to Greenwood, December 17, 1912, KPP, box 915.

106. Greenwood, "On Methods of Research Available in the Study of Medical Problems," *The Lancet* i (1913): 158.

107. Ibid., p. 163.

108. Ibid.

109. William Osler, *The Principles and Practice of Medicine*, rev. 7th ed. (London: D. Appleton and Co., 1909), p. 356.

110. Wright, *Studies in Immunisation*, 2nd series (London: William Heinemann, 1944), p. 243.

111. Ibid., pp. 251–52.

112. George Bernard Shaw, *The Doctor's Dilemma, Getting Married, & The Shewing-Up of Blanco Posnet*, rev. standard ed. (London: Constable, 1932), p. 26. I would like to thank Robert W. Prichard for this reference to the work of Shaw.

113. Ibid., pp. 58–59.

CHAPTER SIX
THE BIRTH OF THE MODERN CLINICAL TRIAL: THE CENTRAL
ROLE OF THE MEDICAL RESEARCH COUNCIL

1. Major Greenwood, "On Methods of Research Available in the Study of Medical Problems: With Special Reference to Sir Almroth Wright's Recent Utterances," *The Lancet* i (1913): 164.

2. George Udny Yule and Major Greenwood, "The Statistics of Anti-Typhoid and Anti-Cholera Inoculations, and the Interpretation of Such Statistics in General," *Proceedings of the Royal Society of Medicine* 8 (1915): 113–90.

3. Yule and Greenwood, "On the Statistical Interpretation of Some Bacteriological Methods Employed in Water Analysis," *Journal of Hygiene* 16 (1917–1918): 40.

4. Major Greenwood to Karl Pearson, February 3, 1911, KPP, box 707.

5. Greenwood to Pearson, February 11, 1911, KPP, box 707.

6. Greenwood to Pearson, February 3, 1911, KPP, box 707.

7. Greenwood to Pearson, March 13 and 27, 1912, KPP, box 707.

8. See Karl Pearson to Raymond Pearl, September 5, 1902, Pearson correspondence, RPP.

9. Pearson to Francis Galton, May 2, 1909, FGP, box 293K.

10. Pearl to Pearson, February 15, 1910, Pearson correspondence, RPP.

11. Quoted in Donald A. MacKenzie, *Statistics in Britain, 1865–1930: The Social Construction of Scientific Knowledge* (Edinburgh: Edinburgh University Press, 1981), p. 111.

12. Joan Austoker and Linda Bryder, "The National Institute for Medical Research and related activities of the MRC," in Austoker and Bryder, eds., *Historical Perspectives on the Role of the MRC: Essays in the History of the Medical Research Council of the United Kingdom and Its Predecessor, the Medical Research Committee, 1913–1953* (Oxford: Oxford University Press, 1989), pp. 35–36.

13. Greenwood to Pearl, May 3, 1921, Pearson correspondence, RPP.

14. Greenwood to Pearson, August 13, 1920, KPP, box 707.

15. MacKenzie, *Statistics in Britain*, p. 117.

16. Pearson to Greenwood, August 14, 1920, KPP, box 915.

17. Greenwood to Pearson, August 16, 1920, KPP, box 707.

18. Greenwood to Pearl, Greenwood correspondence, October 14, 1921, RPP.

19. Greenwood to Pearl, November 23, 1921, RPP.

20. Eugene S. Kilgore, "Relation of Quantitative Methods to the Advance of Medical Science," *Journal of the American Medical Association* (July 10, 1920): 88.

21. Pearl, "Modern Methods in Handling Hospital Statistics," *The Johns Hopkins Hospital Bulletin* 32 (1921): 185.

22. Greenwood to Pearson, December 13, 1924, KPP, box 707.

23. Pearl, *Introduction to Medical Biometry and Statistics* (Philadelphia: Saunders, 1923), p. 21.

24. Yule to Pearl, November 8, 1923; Greenwood to Pearl, November 18, 1923; Pearson to Pearl, December 8, 1930, all in RPP.

25. Greenwood to Pearl, July 7, 1923, Greenwood correspondence, RPP.

26. Greenwood to Pearl, January 25 and February 26, 1923, Greenwood correspondence, RPP.

27. Pearl to Greenwood, April 25, 1927, Greenwood correspondence, RPP.

28. Greenwood to Pearl, March 27, 1923; Pearl to Greenwood, April 9, 1923, Greenwood correspondence, RPP.

29. Greenwood to Pearl, April 14, 1927, Greenwood correspondence, RPP.

30. Greenwood to Pearl, October 2, 1924, Greenwood correspondence, RPP.

31. Thomas B. Turner, *Heritage of Excellence: The Johns Hopkins Medical Institutions, 1914–1947* (Baltimore: The Johns Hopkins University Press, 1974), p. 359.

32. Pearl to Yule, March 28, 1925, Yule correspondence, RPP.

33. Pearl to Greenwood, October 17, 1923, Greenwood correspondence, RPP.

34. Greenwood to Pearl, June 9, 1925, Greenwood correspondence, RPP.

35. Greenwood to Pearl, February 27, 1924, Greenwood correspondence, RPP.

36. Greenwood, *Epidemics and Crowd-Diseases: An Introduction to the Study of Epidemiology* (New York: Macmillan, 1935), pp. 64–65.

37. Greenwood, *The Medical Dictator and Other Biographical Studies* (London: Williams and Norgate, 1936), p. 133.

38. Ibid., p. 136.

39. Ibid., pp. 139, 141.

40. Pearson, "Memorandum on the History, Finance and Present Scheme of Reorganisation of the Galton and Biometric Laboratories," University of London, KPP.

41. "Report of the Professional Board on the Resignation of the Galton Chair by Professor Karl Pearson," University of London, University College, KPP.

42. I. D. Hill, "Austin Bradford Hill—Ancestry and Early Life," *Statistics in Medicine* 1 (1982): 299–300; Sir Harold Himsworth, "Bradford Hill and Statistics in Medicine," *Statistics In Medicine* 1 (1982): 301–2.

43. Editor of *The Lancet*, "Foreword to First Edition," in Sir Austin Bradford Hill, *Principles of Medical Statistics*, 8th ed. (New York: Oxford University Press, 1966), p. iii.

44. Ibid.

45. Hill, "Preface to First Edition," in *Principles of Medical Statistics*, p. vii.

46. Harold M. Schoolman, "The Clinician and the Statistician," *Statistics in Medicine* 1 (1982): 311–12 (documented in the Bibliography under *Statistics in Medicine*).

47. Fisher and his relationship to medicine are discussed in Harry M.

Marks, "Ideas as Reforms: Therapeutic Experiments and Medical Practice, 1900–1980," (Ph.D. dissertation, Massachusetts Institute of Technology, 1987), pp. 158–64.

48. Greenwood, "The Statistician and Medical Research," *British Medical Journal* (1948): 467.

49. Austoker and Bryder, *Historical Perspectives on the Role of the MRC,* pp. 45–52; F.H.K. Green, "The Clinical Evaluation of Remedies," *The Lancet* ii (1954): 1085–90.

50. "Streptomycin Treatment of Pulmonary Tuberculosis: A Medical Research Council Investigation," *British Medical Journal* (1948): 769.

51. Ibid., pp. 769–70.

52. Ibid., pp. 770–71.

53. Ibid., p. 771.

54. Ibid., p. 773.

CHAPTER SEVEN
A. BRADFORD HILL AND THE RISE OF THE CLINICAL TRIAL

1. E. K. Marshall and Margaret Merrell, "Clinical Therapeutic Trial of a New Drug," *Bulletin of The Johns Hopkins Hospital* 85 (1949): 229.

2. Ibid., p. 221.

3. D. D. Reid, "Statistics in Clinical Research," *Annals of the New York Academy of Sciences* 52 (1950): 932. *Note: Documentation for notes 3–10 can be found under the title "The Place of Statistical Methods in Biological and Chemical Experimentation" in the Primary Sources section of the Bibliography.*

4. Ibid., p. 933.

5. Ibid., p. 933.

6. Ibid., p. 934.

7. Donald Mainland, "Statistics in Clinical Research: Some General Principles," *Annals of the New York Academy of Sciences* 52 (1950): 925.

8. Ibid., p. 927.

9. Ibid., p. 929.

10. Ibid., p. 929.

11. Sir Austin Bradford Hill, "The Clinical Trial—III," in *Statistical Methods in Clinical and Preventive Medicine* (London: E. and S. Livingstone, 1962), p. 29. Rpt. (Hill, "The Clinical Trial") from the *New England Journal of Medicine* 247 (July 24, 1952): 113.

12. Hill, "The Clinical Trial—III," p. 31.

13. Ibid., p. 42.

14. For a survey of this debate, see H. Marks, "Ideas as Reforms: Therapeutic Experiments and Medical Practice, 1900–1980" (Ph.D. dissertation, Massachusetts Institute of Technology, 1987), p. 164.

15. L. J. Witts, Introduction to *Medical Surveys and Clinical Trials,* 2nd ed., ed. L. J. Witts (London: Oxford University Press, 1964), p. 5.

16. Foreword to (Council for International Organizations of Medical Sci-

ences) *Controlled Clinical Trials* (Oxford: Blackwell Scientific Publications, 1960), p. vii.

17. Hill, "Aims and Ethics," in ibid., p. 3.

18. Ibid., p. 4.

19. Ibid., p. 6.

20. L. J. Witts, "The Ethics of Controlled Clinical Trials," in ibid., p. 11.

21. Ibid., pp. 12–13.

22. Sir George W. Pickering, "Conclusion: The Physician," in ibid., pp. 164–65.

23. Ibid., p. 165.

24. The fact that the clinical trial was largely imposed on the medical profession by the government as a result of public outcry may suggest interesting parallels to the Clinton administration's present efforts at health care reform in the United States.

25. Such a social interpretation of the rise of statistical knowledge has been argued for vigorously by Theodore M. Porter, "The Quantification of Uncertainty After 1700: Statistics Socially Constructed?" in George M. von Furstenberg, ed., *Acting Under Uncertainty: Multidisciplinary Conceptions* (Dordrecht: Kluwer, 1990), esp. pp. 62–71.

26. For further account of the details behind this law, see William J. Curran, "Governmental Regulation of the Use of Human Subjects in Medical Research: The Approach of Two Federal Agencies," in *Daedalus* (Spring, 1969), pp. 549–51.

27. Ibid., pp. 551–52.

28. Ibid., p. 552.

29. Ibid., pp. 555–56.

30. For a comparison of how the double-blind procedure triumphed in the United States and Western Europe, see Henk J.H.W. Bodewitz, Henk Buurma, and Gerard H. de Vries, "Regulatory Science and the Social Management of Trust in Medicine," in Wiebe E. Bijker, Thomas P. Hughes, and Trevor Pinch, eds., *The Social Construction of Technological Systems* (Cambridge: MIT Press, 1987), esp. pp. 252–57. A comparison of the United States and Great Britain can also be found in Stuart J. Pocock, *Clinical Trials: A Practical Approach* (New York: Wiley, 1983), pp. 26–27.

31. This issue is the theme of chapter 7 in Gerd Gigerenzer et al., *The Empire of Chance: How Probability Changed Science and Everyday Life* (Cambridge: Cambridge University Press, 1989), pp. 235–70.

Conclusion

1. Sir Richard Doll, "Clinical Trials: Retrospect and Prospect," *Statistics in Medicine* 1 (1982): 343.

2. Some of the contemporary debates regarding these issues are outlined in the recent articles by Valerie Miké. See "Understanding Uncertainties in Medical Evidence: Professional and Public Responsibilities," in Deborah G. Mayo and Rachelle D. Hollander, eds., *Acceptable Evidence: Science*

and Values in Risk Management (Oxford: Oxford University Press, 1991), pp. 115–36 and "Philosophers Assess Randomized Clinical Trials: The Need for Dialogue," *Controlled Clinical Trials* 10 (1989): 244–53.

3. These contrasting visions of science are surveyed from a contemporary perspective in Kenneth F. Schaffner, "Causing Harm: Epidemiological and Physiological Concepts of Causation," in Mayo and Hollander, *Acceptable Evidence*, pp. 204–17.

4. The inclusion of women and women's diseases in clinical trials has been advocated in the American political context by such notable figures as U. S. Representative Patricia Schroeder and the current first lady Hillary Rodham Clinton.

5. "A Critique of Pure Reason, A Passion to Survive," *New York Times*, 21 October, 1990, p. E4.

6. The Clinton administration has allowed trials to begin on this drug in the United States in the fall of 1994. See "RU-486 Is on Its Way to the U.S.," *The Washington Post*, 17 May, 1994, pp. A1, A4.

7. David A. Hollinger has characterized the difference between modernism and postmodernism in terms of this attack on universalism in his recent article "How Wide the Circle of the 'We'? American Intellectuals and the Problem of the Ethnos since World War II," *American Historical Review* 98 (April 1993): 317–37; see esp. pp. 321–22.

8. Allan Megill, "Four Senses of Objectivity," in "Rethinking Objectivity, I," *Annals of Scholarship* 8 (1991): 301–20.

9. See Michael Polanyi, *Personal Knowledge* (Chicago: University of Chicago Press, 1957); Steve Fuller has chosen to characterize the belief that science has a nonverbal tacit dimension as "deep science" and the belief that science is a verbal craft as "shallow science" in his new book *Philosophy, Rhetoric, and the End of Knowledge: The Coming of Science and Technology Studies* (Madison: University of Wisconsin Press, 1993), pp. 11–12 and *passim*.

10. Most of the scholarly literature tends to emphasize (as has this study in the analysis of Bernard) the tensions between the clinician and the laboratory researcher rather than their points of agreement. See Gerald L. Gieson, "Divided We Stand: Physiologists and Clinicians in the American Context," and Russell C. Maulitz, "'Physician versus Bacteriologist': The Ideology of Science in Clinical Medicine," both in Morris J. Vogel and Charles E. Rosenberg, eds., *The Therapeutic Revolution: Essays in the Social History of American Medicine* (Philadelphia: University of Pennsylvania Press, 1979), pp. 67–90 and 91–107, respectively.

11. P.-C.-A. Louis, *Researches on the Effects of Bloodletting*, trans. C. G. Putnam (Hilliard, Gray, 1836), p. 68.

12. Gustav Radicke, "On the Deduction of Physiological and Pharmaco-Dynamical Probabilities from Co-ordinated Series of Observations," trans. Francis T. Bond (London: The New Sydenham Society, 1861), p. 270.

13. Major Greenwood, "On Methods of Research Available in the Study of Medical Problems," *The Lancet* i (1913): 163.

14. Sir Austin Bradford Hill, "The Philosophy of the Clinical Trial," in his

collection of essays entitled *Statistical Methods in Clinical and Preventive Medicine* (London: E. and S. Livingstone, 1962), p. 13.

15. On the role of quantification as a form of public rhetoric, see Theodore M. Porter, "Objectivity and Authority: How French Engineers Reduced Public Utility to Numbers," *Poetics Today* 12 (Summer 1991): 245–65. See esp. pp. 248–52.

16. Theodore M. Porter has argued very cogently for this point in "Objectivity as Standardization: The Rhetoric of Impersonality in Measurement, Statistics and Cost-Benefit Analysis," *Annals of Scholarship* 9 (1992): 19–59; see esp. pp. 28–32.

17. Allan M. Brandt cited the historical study of the changing nature of the doctor-patient relationship as an especially prominent feature of the "new history of medicine," in "Emerging Themes in the History of Medicine," *The Milbank Quarterly* 69 (1991): 199–214; see esp. 207–9.

18. Claude Bernard, *An Introduction to the Study of Experimental Medicine*, trans. Henry C. Greene (New York: Dover, 1957), p. 136.

19. Carl Auguste Wunderlich, *On the Temperature in Diseases: A Manual of Medical Thermometry*, trans. W. Bathurst Woodman (London: New Sydenham Society, 1871 [1868]), pp. 48–49.

20. See Stanley Joel Reiser, *Medicine and the Reign of Technology* (Cambridge: Cambridge University Press, 1978), p. 117.

21. See Anne Hunsaker Hawkins, "Oliver Sack's *Awakenings*: Reshaping Clinical Discourse," *Configurations: A Journal of Literature, Science, and Technology* 1 (1993): 235.

22. Examples of recent scholarly works that put forward such claims include Kathryn Montgomery Hunter, *Doctors' Stories: The Narrative Structure of Medical Knowledge* (Princeton: Princeton University Press, 1991), and George Khushf, "Post-Modern Reflections on the Ethics of Naming," in José Luis Peset and Diego Gracia, eds., *The Ethics of Diagnosis* (Dordrecht: Kluwer, 1992), pp. 275–300.

23. The same could not be said about public health statistics in which nineteenth-century physicians played a prominent role. Individuals such as William Farr and William Guy were, in fact, in the vanguard of nineteenth-century statistical thinking on the collection of social facts.

24. See Charles E. Rosenberg and Janet Golden, eds., *Framing Disease: Studies in Cultural History* (New Brunswick: Rutgers University Press, 1992), esp. part 3, "Negotiating Disease: The Public Arena."

25. Contrasting attitudes to the value of the computer in diagnosis are surveyed in Edmund D. Pellegrino, "Value Disiderata in the Logical Structuring of Computer Diagnosis," Henrik R. Wulff, "Computers and Clinical Thinking," and Edmond A. Murphy, "Critique of Diagnostic Formalism," all in José Luis Peset and Diego Gracia, eds., *The Ethics of Diagnosis*, pp. 173–95, 243–54, and 255–67, respectively.

26. The phrase was made famous by Clifford Geertz in his *The Interpretation of Cultures* (New York: Basic Books, 1973), esp. chapter 1: "Thick Description: Toward an Interpretive Theory of Culture," pp. 3–30.

BIBLIOGRAPHY

Note : Seeming inconsistencies, here and in the notes, between the spellings *Medezin* and *Medecin, medezinisch* and *medecinisch,* etc., reflect a change in the conventional German spelling from the nineteenth to the twentieth century.

ARCHIVAL SOURCES

London. University College. Francis Galton Papers (FGP).
Philadelphia, Pennsylvania. American Philosophical Society. Raymond Pearl Papers (RPP).
London. University College. Karl Pearson Papers (KPP).
London. School of Hygiene and Tropical Medicine. Ronald Ross Papers.

PRIMARY SOURCES

Alison, William Pulteney. "Notes on the Application of Statistics to Questions in Medical Science, Particularly as to the External Causes of Diseases." *Report of the BAAS* (1855): 155–59.
Allbutt, Thomas Clifford. "An Introductory Address on Theory and Practice." *The Lancet* ii (1899): 925–28.
———, ed. *A System of Medicine by Many Writers.* New York: Macmillan, 1901.
"Antityphoid Inoculation." *British Medical Journal* (1904): 1259–61.
"Anti-Typhoid Inoculation in the Army," *Times* (London), 24 January 1904, p. 7.
"The Application of Mathematical Formulae to Medical Statistics." *British Medical Journal* (1875): 72–73.
Auber, T.-C.-E. *Traité de philosophie médicale, ou exposition des verités générales et fondamentales de la médecine.* Paris: Germer Baillière, 1839.
Bartlett, Elisha. *An Essay on the Philosophy of Medical Science.* Philadelphia: Lea and Blanchard, 1844.
Beneke, F. W. "A Reply to Professor Radicke's Paper, 'On The Importance and Value of Arithmetic Means.'" Trans. Francis T. Bond. London: New Sydenham Society, 1861.
Bernard, Claude. "De l'emploi des moyennes en physiologie expérimentale à propos de l'influence de l'effeuillage des betteraves sur la production." *Comptes rendus de l'Académie des Sciences* 81 (1875): 698–703.
———. *Principes de médecine expérimentale.* Paris: Presses Universitaires de France, 1947.
———. *An Introduction to the Study of Experimental Medicine.* Trans. Henry Copley Greene. New York: Dover, 1957.

Bienaymé, Irenée Jules. "Calcul des probabilités: Applications à la statistique médicale." *Procès verbaux de la Société Philomatique de Paris* (1840): 10–13.

———. "Considérations à l'appui de la découverte de Laplace sur la loi de probabilité dans la méthode des moindres carrés." *Comtes rendus des séances de l'Académie des Sciences* 37 (1853): 309–34.

Billings, John S. "Our Medical Literature." *Lancet* ii (1881): 265–70.

Bouillaud, Jean-Baptiste. *Essai sur la philosophie médicale.* Paris: Libraire des sciences médicales, 1836.

Broussais, Casimir. *De la statistique appliquée à la pathologie et à la thérapeutique.* Paris: J.-B. Baillière, Libraire de l'Académie Royale de Médecine, 1840.

Bulletin de l'Académie Royale de Médecine 1 (1836/37): 5–8, 605–806. [Academy of Medicine debate]

Cabanis, Pierre-Jean-Georges. *Oeuvres philosophiques de Cabanis.* Vol. 1. Paris: Presses Universitaires de France, 1956.

Cheever, David W. "The Value and the Fallacy of Statistics in the Observation of Disease." *The Boston Medical and Surgical Journal* 63 (1861): 449ff.; 476ff.; 496ff.; 512ff.; 534ff.

Civiale, Jean. *Traité de l'affection calculeuse, ou recherches sur la formation, les caractères physiques et chimiques, les causes, les signes, et les effets pathologiques de la pierre et de la gravelle suives d'un Essai de statistique sur cette maladie.* Paris: Crochard et Comp, 1838.

Comte, Auguste. *Cours de philosophie positive.* 2nd ed. Paris: J. B. Baillière, 1864.

Condorcet, M.-J.-A.-N. *Condorcet: Selected Writings.* Edited, with an introduction by Keith Michael Baker. Indianapolis: Bobbs-Merrill, 1976.

Controlled Clinical Trials. Oxford: Blackwell Scientific Publications, 1960.

The Cyclopedia of Anatomy and Physiology. Ed. Robert B. Todd. London: Longman, Brown, Green, Longmans and Roberts, 1849–52. S.v. "Statistics, medical."

Danvin, Bruno. "De la methode numérique et de ses avantages dans l'étude de la médecine." *Thèses de l'Ecole de Médecine.* Paris, 1831.

Dechambre, Amédée. "La Statistique médicale." *Gazette hebdomadaire de médecine et de chirurgie* 11 (1874): 195–96, 209–12.

Dictionnaire encyclopédique des sciences médicales. Ed. A. Dechambre, L. Hahn, et al. Paris: P. Asselin and G. Masson, 1883. S.v. "Statistique médicale—Application à la médecine," "Moyenne."

Dictionnaire de médecine, ou repertoire général des sciences médicales considerées sous le rapport théorique et pratique. 2nd ed. Ed. N.-P. Adelon et al. Paris: Becket jne, 1782–1862. S. v. "Statistique, médicale."

Double, François-Joseph. "Recherches de statistique sur l'affection calculeuse." *Comtes rendus de l'Académie des Sciences* 1 (1835): 167–77.

———. "Statistique appliquée à la médecine." *Comptes rendus de l'Académie des Sciences* 1 (1835): 280–81.

———. Introduction. *Traité de médecine-pratique de Jean-Pierre Frank.* Paris: J.-B. Baillière Libraire de l'Académie Royale de Médecine, 1842.

Fick, Adolf. *Die Medicinische Physik.* Braunschweig: Friedrich Vieweg, 1866.

Fisher, R. A. The Design of Experiments. 8th ed. Edinburgh: Oliver and Boyd, 1966 [1935].

Flint, Austin. "Remarks on the Numerical System of Louis." *New York Journal of Medicine and Science* 4 (1841): 283–303.

Gavarret, Jules. "Application de la statistique à la médecine: Réponse à l'examen critique auquel M. le docteur Valleix a soumis, dans le numéro de mai 1840 des *Archives générales de médecine*, l'ouvrage de M. Gavarret intitulé Principes généraux de statistique médicale ou développement des règles qui doivent présider à son emploi." *L'Expérience* 6 (1840): 1–12.

———. *Principes généraux de statistique médicale.* Paris: Libraires de la Faculté de Médecine de Paris, 1840.

Green, F.H.K. "The Clinical Evaluation of Remedies." *The Lancet* ii (1954): 1085–90.

Greenwood, Major. "Statistical Considerations Relative to the Opsonic Index." *The Practitioner* 80 (1908): 641–46.

———. "A Statistical View of the Opsonic Index." *Proceedings of the Royal Society of Medicine* 2 (1909): 145–58.

———. "On Methods of Research Available in the Study of Medical Problems: With Special Reference to Sir Almroth Wright's Recent Utterances," *The Lancet* i (1913): 158–65.

———. *Epidemics and Crowd-Diseases: An Introduction to the Study of Epidemiology.* New York: Macmillan, 1935.

———. *The Medical Dictator and Other Biographical Studies.* London: Williams and Norgate, 1936.

———. "The Statistician and Medical Research." *British Medical Journal* (1948): 467–68.

Griffin, William, and Daniel Griffin. *Medical and Physiological Problems; being chiefly researches for Correct Principles of Treatment in Disputed Points of Medical Practice.* London: Sherwood, Gilbert and Piper, 1843.

Guy, William A. "On the Value of the Numerical Method as Applied to Science, but Especially to Physiology and Medicine." *Journal of the Statistical Society of London* 2 (1839): 25–47.

———. "The Numerical Method, and Its Application to the Science and Art of Medicine." *British Medical Journal* (1860): 331ff., 371ff., 409ff., 467ff., 553ff., 596ff.

———. "On the Original and Acquired Meaning of the Term 'Statistics,' and on the Proper Functions of a Statistical Society." *Journal of the Statistical Society* 28 (1865): 478–93.

Harvey, W. F., and Anderson McKendrick. "The Opsonic Index—A Medico-Statistical Enquiry." *Biometrika* 7 (1909-1910): 64–95.

Herschel, John. "Letters Addressed to H.R.H. the Grand Duke of Saxe-Coburg and Gotha on the Theory of Probabilities as Applied to the Moral and Political Sciences." *Edinburgh Review* 92 (1850): 1–57.

Hill, Sir Austin Bradford. *Statistical Methods in Clinical and Preventive Med-*

icine. London: E. and S. Livingstone, 1962. Rpt. from (Hill, "The Clinical Trial") the *New England Journal of Medicine* 247 (1952): 113–19.

Hill, Sir Austin Bradford. *Principles of Medical Statistics*. 8th ed. New York: Oxford University Press, 1966.

Hirschberg, Julius. *Die mathematischen Grundlagen der medizinischen Statistik, elementar Dargestellt*. Leipzig: Veit, 1874.

Jessen, Willers. "Zur Analytischen Statistik." *Zeitschrift für Biologie* 3 (1867): 128–36.

Kilgore, Eugene S. "Relation of Quantitative Methods to the Advance of Medical Science." *Journal of the American Medical Association* (July 10, 1920): 86–88.

Lacroix, Sylvestre François. *Traité élémentaire du calcul des probabilités*. Paris: Bachelier, 1833.

Laplace, Pierre-Simon. *A Philosophical Essay on Probabilities*. 6th ed. Trans. Frederick Wilson Truscott and Frederick Lincoln Emory. New York: Dover, 1952.

Liebermeister, Carl. "Ueber Wahrscheinlichkeitsrechnung in Anwendung auf therapeutische Statistik." *Volkmanns Sammlung klinischer Vorträge*, no. 110: Innere Medicin, No. 39 (1877): 935–62.

Lister, Joseph. "Effects of the Antiseptic System of Treatment upon the Salubrity of a Surgical Hospital." *The Lancet* i (1870): 4–6; 40–42.

———. "Further Evidence Regarding the Effects of the Antiseptic System of Treatment upon the Salubrity of a Surgical Hospital." *The Lancet* ii (1870): 287–89.

Louis, Pierre-Charles-Alexandre. *Anatomical, Pathological and Therapeutic Researches upon the Diease Known under the Name of Gastro-Enterite Putrid, Adynamic, Ataxic, or Typhoid Fever, etc., Compared with the Most Common Acute Diseases*, 2 vols. Trans. Henry I. Bowditch. Boston: Hilliard, Gray, 1836 [vol. 1]. Boston: Isaac R. Butts, 1836 [vol. 2].

———. *Pathological Researches on Pthisis*. Trans. Charles Cowan. Boston: Hilliard, Gray, 1836.

———. *Researches on the Effects of Bloodletting in Some Inflammatory Diseases, and on the Influence on Tartarized Antimony and Vesication in Pneumonitis*. Trans. C. G. Putnam. Boston: Hilliard, Gray, 1836.

Marshall, E. K., and Margaret Merrill. "Clinical Therapeutic Trial of a New Drug." *Bulletin of the Johns Hopkins Hospital* 85 (1949): 221–30.

Martius, Friedrich. "Die Principien der wissenschaftlichen Forschung in der Therapie." *Volkmanns Sammlung klinischer Vorträge* no. 139: Innere Medicin, no. 47 (1878): 1169–88.

———. "Die Numerische Methode (Statistik und Wahrscheinlichkeitsrechnung) mit besonderer Berücksichtigung ihrer Anwendung auf die Medicin." *Virchows Archiv für pathologische Anatomie und Physiologie und für klinische Medicin*, 83 (1881): 336–77.

———. "Autobiography." In *Die Medizin der Gegenwart in Selbstdarstellungen*. Ed. L. R. Grote. Leipzig: Felix Meiner, 1923.

"Mathematics and Medicine." *British Medical Journal* (1911): 449.

"The Medicine of the Future." *British Medical Journal* (1907): 333–34.

Mémoires de la Société Médicale d'Observation, 3 vols. Paris: 1837, 1844, and 1856.

"Miscellanea." *Journal of the Statistical Society* 40 (1877): 468–77.

Navier, Claude. "Statistique appliquée à la médecine." *Comptes rendus de l'Académie des Sciences* 1 (1835): 247–50.

Oesterlen, Friedrich. *Medical Logic*. Trans. G. Whitley. London: Sydenham Society, 1855 [1852].

———. *Handbuch der Medicinischen Statistik*. Tübingen: H. Laupp'schen, 1874.

Osler, William. *The Principles and Practice of Medicine*. Rev. 7th ed. London: D. Appleton, 1909.

Pearl, Raymond. "Modern Methods in Handling Hospital Statistics." *The Johns Hopkins Hospital Bulletin* 32 (1921): 184–94.

———. *Introduction to Medical Biometry and Statistics*. Philadelphia: Saunders, 1923.

Pearson, Karl. "Antityphoid Inoculation." *British Medical Journal* (1904): 1,432, 1,542, 1,667, 1,775–76.

———. "Report on Certain Enteric Fever Inoculation Statistics." *British Medical Journal* (1904): 1,243–46.

———. *The Grammar of Science*. 3rd ed. New York: Macmillan, 1911.

———. "The Opsonic Index—'Mathematical Error and Functional Error.'" *Biometrika* 8 (1911-1912): 203–24.

Pinel, Philippe. "Resultats d'observations et constructions des tables pour servir à determiner le degré de probabilité de la guérison des aliénés." *Institut de France, Mémoires classe des sciences mathématique et physique* 8 (1807): 169–205.

———. *Traité medico-philosophique sur l'aliénation mentale*, 2nd ed. Paris, 1809.

"The Place of Statistical Methods in Biological and Chemical Experimentation." *Annals of the New York Academy of Sciences* 52 (March 10, 1950): 789–942.

Poisson, Siméon-Denis. *Recherches sur la probabilité des jugements en matière criminelle et en matière civile*. Paris: Bachelier, 1837.

Pye-Smith, Philip Henry. "Medicine as a Science and Medicine as an Art." *The Lancet* ii (1900): 309–12.

Quetelet, Lambert Adolphe Jacques. *Letters Addressed to H.R.H. The Grand Duke of Saxe Coburg and Gotha*. Trans. Olinthus Gregory Downes. London: Charles and Edwin Layton, 1849.

———. *A Treatise On Man and the Development of His Faculties*. Trans. R. Knox. New York: Burt Franklin, Research Source Works Series #247, 1962 [1835].

Radicke, Gustav. "On the Deduction of Physiological and Pharmaco-Dynamical Probabilities from Co-ordinated Series of Observations." Trans. Francis T. Bond. London: The New Sydenham Society, 1861.

———. "On the Importance and Value of Arithmetic Means." Trans. Francis T. Bond. London: The New Sydenham Society, 1861.

"Research in Medicine." *The Lancet* ii (1905): 771–72.

"Review of 'Principes généraux de statistique médicale, ou développement des règles qui doivent présider à son emploi.'" *British and Foreign Medical Review* 12 (July, 1841): 1–21.

"Revue analytique et critique des Journaux de médecine français." *Revue médicale française et etrangère, Journal des progres de la médecine Hippocratique* 2 (1840): 243–46.

Risueño d'Amador, Benigno Juan Isidoro R. *Mémoire sur le calcul des probabilités appliqué à la médecine.* Paris, 1837.

Schweig, G. "Auseinandersetzung der statistischen Methode in besondere Hinblick auf das medicinische Bedürfniss" *Archiv für physiologische Heilkunde* 13 (1854): 305–55.

Seidel, Ludwig. "Ueber den numerischen Zusammenhang welcher zwischen der Häufigkeit der Typhus-Erkrankungen und dem Stande des Grundwassers während der letzten 9 Jahre in München hervorgetreten ist." *Zeitschrift für Biologie* 1 (1865): 221–36.

Shaw, George Bernard. *The Doctor's Dilemma, Getting Married, and The Shewing of Blanco Posnet.* Rev. standard ed. London: Constable, 1932.

"Streptomycin Treatment of Pulmonary Tuberculosis: A Medical Research Council Investigation." *British Medical Journal* (1948): 769–82.

Trousseau, Armand. *Lectures on Clinical Medicine, Delivered at the Hôtel-Dieu, Paris.* 3rd ed. Trans. John Rose Cormack. London: The New Sydenham Society, 1869.

"Vaccine Therapy—Its Administration, Value, and Limitations: A Discussion." *Proceedings of the Royal Society of Medicine* 3 (May 23, May 25, June 1, June 8, June 15, June 22, 1910): 1–216.

Valleix, F.-L.-I. "De l'application de la statistique à la médecine: Examen critique de l'ouvrage de M. Gavarret." *Archives générales de médecine* 8 (1840): 5–39.

Vierordt, Karl. "Notes on Medical Statistics." Trans. Francis T. Bond. London: The New Sydenham Society, 1861.

White, J.D.C., and Major Greenwood. "A Biometric Study of Phagocytosis with Special Reference to the 'Opsonic Index.'" *Biometrika* 6 (1908-1909): 376–401.

———. "A Biometric Study of Phagocytosis with Special Reference to the 'Opsonic Index,' Second Memoir. On the Distribution of the Means of Samples." *Biometrika* 7 (1909-1910): 505–30.

Witts, L. J., ed. *Medical Surveys and Clinical Trials.* 2nd ed. London: Oxford University Press, 1964.

Wright, Almroth Edward. "Antityphoid Inoculation." *British Medical Journal* (1904): 1,343–45, 1,489–91, 1,614, 1,727.

———. *A Short Treatise on Anti-Typhoid Inoculation.* Westminster: Archibald Constable, 1904.

———. "On Some Points in Connection with Vaccine-Therapy and Therapeutic Immunisation Generally." *The Practitioner* 80 (1908): 565–606.

———. *Studies in Immunisation,* 2nd series. London: William Heinemann, 1944.

Wunderlich, Carl Auguste. "Das Verhältniss der physiologischen Medicin zur ärztlichen Praxis." *Archiv für physiologische Heilkunde* 4 (1845): 1–13.

———. *On the Temperature in Diseases: A Manual of Medical Thermometry.* Trans. W. Bathurst Woodman. London: New Sydenham Society, 1871 [1868].

Wunderlich, Carl, and A. Roser. "Einleitung: Über die Mangel der heutigen deutschen Medicin und über die Nothwendigkeit einer entschieden wissenschaftlichen Richtung in derselben." *Archiv für physiologische Heilkunde* 1 (1842): 1–16.

Yule, George Udny, and Major Greenwood. "The Statistics of Anti-Typhoid and Anti-Cholera Inoculations, and the Interpretation of Such Statis tics in General." *Proceedings of the Royal Society of Medicine* 8 (1915): 113–90.

———."On the Statistical Interpretation of Some Bacteriological Methods Employed in Water Analysis." *Journal of Hygiene* 16 (1917–18): 36–54.

SECONDARY SOURCES

Ackerknecht, Erwin H. "Elisha Bartlett and the Philosophy of the Paris Clinical School." *Bulletin of the History of Medicine* 24 (1950): 43–60.

———. "Broussais, or A Forgotten Medical Revolution." *Bulletin of the History of Medicine* 27 (1953): 322–33.

———. *Medicine at the Paris Hospital, 1794–1848.* Baltimore: The Johns Hopkins University Press, 1967.

Albury, W. R. "Heart of Darkness: J. N. Corvisart and the Medicalization of Life." *Historical Reflections* 9 (1982): 17–31.

Austoker, Joan, and Linda Bryder, eds. *Historical Perspectives on the Role of the MRC: Essays in the History of the Medical Research Council of the United Kingdom and of Its Predecessor, the Medical Research Committee, 1913–1953.* Oxford: Oxford University Press, 1989.

Baker, Keith Michael. *Condorcet: From Natural Philosophy to Social Mathematics.* Chicago: University of Chicago Press, 1975.

Bariety. "Louis et la methode numérique." *Clio Medica* 7 (1972): 177–83.

Ben-David, Joseph. *The Scientist's Role in Society: A Comparative Study.* Englewood Cliffs: Prentice Hall, 1971.

Bijker, Wiebe E., Thomas P. Hughes, and Trevor Pinch, eds. *The Social Construction of Technology Systems.* Cambridge: MIT Press, 1987.

Biographisches Lexikon der Hervorragenden Ärtze. 2nd ed. Ed. W. Haberling and H. Vierordt. Berlin and Vienna: Urban and Schwarzenberg, 1932–35. S.v. "Oesterlen, Friedrich"; "Raige-Delorme, Jacques"; "Trousseau, Armand T"; "Valleix, François-Louis-Isidore"; "Wunderlich, Karl Reinhold August W."

Bollet, Alfred Jay. "Pierre Louis: The Numerical Method and the Foundation of Quantitative Medicine." *The American Journal of Medical Science* 266 (1973): 93–101.

Brandt, Allan M. "Emerging Themes in the History of Medicine." *The Milbank Quarterly* 69 (1991): 199–214.

Bull, J. P. "The Historical Development of Clinical Therapeutic Trials." *Journal of Chronic Diseases* 10 (1959): 218–48.

Cannon, Susan Faye. *Science in Culture: The Early Victorian Period.* New York: Science History Publications, 1978.

Cassedy, James H. *American Medicine and Statistical Thinking, 1800–1860.* Cambridge: Harvard University Press, 1984.

Clendening, Logan, ed. *Source Book of Medical History.* New York: Dover, 1960 [1942].

Colebrook, Leonard. *Almroth Wright: Provocative Doctor and Thinker.* London: William Heinemann, 1954.

Coleman, William. *Death is a Social Disease: Public Health and Political Economy in Early Industrial France.* Madison: University of Wisconsin Press, 1982.

———. "Neither Empiricism nor Probability: The Experimental Approach." In *Probability Since 1800: Interdisciplinary Studies of Scientific Development.* Ed. Michael Heidelberger, Lorenz Krüger, and Rosemarie Rheinwald. Bielefeld: Universität Bielefeld, 1983.

———. "The Cognitive Basis of the Discipline: Claude Bernard on Physiology." *Isis* 76 (1985): 49–70.

Cope, Zachary. *Almroth Wright: Founder of Modern Vaccine-Therapy.* London: Thomas Nelson, 1966.

Cowan, Ruth S. "Francis Galton's Statistical Ideas: The Influence of Eugenics." *Isis* 63 (1972): 509–28.

"A Critique of Pure Reason, A Passion to Survive." *New York Times,* 21 October 1990, p. E4.

Cullen, Michael J. *The Statistical Movement in Early Victorian Britain.* New York: Harvester, 1975.

Curran, William J. "Governmental Regulation of the Use of Human Subjects in Medical Research: The Approach of Two Federal Agencies." *Daedalus* (Spring, 1969): 542–94.

Daston, Lorraine J. *Classical Probability in the Enlightenment.* Princeton: Princeton University Press, 1988.

Dictionary of National Biography. Ed. Sir Leslie Stephen and Sir Sidney Lee. 3 vols. Oxford and New York: Oxford University Press, 1992 [1921–27]. S.v. "Guy, William Augustus."

Dictionary of Scientific Biography. Ed. Charles C. Gillispie. New York: Scribner, 1970–. S.v. "Von Petenkoffer, Max Josef."

Dictionnaire de biographie francaise. Ed. R. D'Amat and R. Limouzin-Lamothe. Paris: Letouzey et Ané, 1967. S.v. "Double, François-Joseph" and "Gavarret, Jules."

Dictionnaire universel des contemporains, 4th ed. Ed. G. Vapereau, suppl. L Garnier. Paris: Hachette, 1873. S.v. "Dubois [d'Amiens], Frederick."

English, Peter C. "Therapeutic Strategies to Combat Pneumococcal Disease: Repeated Failure of Physicians to Adopt Pneumococcal Vaccine, 1900–1945." *Perspectives in Biology and Medicine* 30 (1987): 170–85.

Eyler, John M. *Victorian Social Medicine: The Ideas and Methods of William Farr.* Baltimore: The Johns Hopkins University Press, 1979.

Foucault, Michel. *The Birth of the Clinic.* Trans. A. M. Sheridan Smith. New York: Vintage, 1973 [1963].

Fox, Robert, and George Weisz, eds. *The Organization of Science and Technology in France, 1808–1914.* Cambridge: Cambridge University Press, 1980.

Fuller, Steve. *Philosophy, Rhetoric and the End of Knowledge: The Coming of Science and Technology Studies.* Madison: University of Wisconsin Press, 1993.

Geertz, Clifford. *The Interpretation of Cultures.* New York: Basoci, 1973.

Gelfand, Toby. *Professionalizing Modern Medicine: Paris Surgeons and Medical Science and Institutions in the Eighteenth Century.* Westport, Conn.: Greenwood Press, 1980.

Giere, Ronald, and Richard Westfall, eds. *Foundations of the Scientific Method: The Nineteenth Century.* Bloomington: Indiana University Press, 1973.

Gigerenzer, Gerd, Zeno Swijtink, Theodore Porter, Loraine Daston, John Beatty, and Lorenz Krüger. *The Empire of Chance: How Probability Changed Science and Everyday Life.* Cambridge: Cambridge University Press, 1989.

Gillispie, Charles C. *Science and Polity in France at the End of the Old Regime.* Princeton: Princeton University Press, 1980.

Goldman, Lawrence. "The Origins of British 'Social Science': Political Economy, Natural Science, and Statistics, 1830–1835." *Historical Journal* 26 (1983): 594–609.

———. "Statistics and the Science of Society in Early Victorian Britain; An Intellectual Context for the General Register Office." *Social History of Medicine* 4 (1991): 415–34.

Goldstein, Jan. *Console and Classify: The French Psychiatric Profession in the Nineteenth Century.* Cambridge: Cambridge University Press, 1987.

Grattan-Guinness, Ivor. "Essay Review: Emergences of Probability?" *Annals of Science* 45 (1988): 643–46.

Hacking, Ian. *The Emergence of Probability.* Cambridge: Cambridge University Press, 1975.

———. "Nineteenth Century Cracks in the Concept of Determinism." *Journal of the History of Ideas* 44 (1983): 455–75.

———. *The Taming of Chance.* Cambridge: Cambridge University Press, 1990.

Hahn, Roger. "Laplace as a Newtonian Scientist." Los Angeles: William Andrews Clark Memorial Library, 1967.

Hawkins, Anne Hunsaker. "Oliver Sack's *Awakenings*: Reshaping Clinical Discourse." *Configurations: A Journal of Literature, Science, and Technology* 1 (1993): 229–45.

Heubner, O. "C. A. Wunderlich: Nekrolog," *Archiv der Heilkunde* 19 (1878): 289–320.

Heyde, C. C., and E. Seneta. *I. J. Bienaymé: Statistical Theory Anticipated.* New York: Springer-Verlag, 1977.

Hilts, Victor L. "*Aliis Exterendum,* or, The Origins of the Statistical Society of London." *Isis* 69 (1978): 21–43.

Hogben, Lancelot. "Major Greenwod, 1880–1949." *Obituary Notices of Fellows of the Royal Society* (1950): 139–41.

Hollinger, David A. "How Wide the Circle of the 'We'? American Intellectuals and the Problem of the Ethnos since World War II." *American Historical Review* 98 (April 1993): 317–37.

Holmes, Frederick Laurence, and William Coleman, eds. *The Investigative Enterprise: Experimental Physiology in Nineteeenth-Century Medicine.* Berkeley: University of California Press, 1988.

Hunt, Lynn, ed. *The New Cultural History.* Berkeley: University of California Press, 1989.

Hunter, Kathryn Montgomery. *Doctor's Stories: The Narrative Structure of Medical Knowledge.* Princeton: Princeton University Press, 1991.

Huron, Roger. "La Statistique médicale en France à l'époque romantique." *Mémoires des Sciences, Inscriptions et Belles-Lettres de Toulouse* 138 (1976): 121–39.

Jokl, Alexander. "Julius Hirschberg." *American Journal of Ophthalmology* 48 (1959): 329–39.

Kevles, Daniel J. *In the Name of Eugenics: Genetics and the Uses of Human Heredity.* Berkeley: University of California Press, 1985.

Koelbing, Huldrych M. "Carl Liebermeister (1833–1901), der erst Chefarzt der Basler medizinischen Universitätsklinik." *Gesnerus* 26 (1969): 233–48.

Krüger, Lorenz, Lorraine J. Daston, Michael Heidelberger, Gerd Gigerenzer, and Mary S. Morgan, eds. *The Probabilistic Revolution.* 2 vols. Cambridge: MIT Press, 1987.

Laborde. "Bulletin hebdomadaire: Le Professeur Gavarret," *La Tribune médicale,* 2nd series, 22 (1890): 577–80.

Latour, Bruno. *Science in Action: How to Follow Scientists and Engineers through Society.* Cambridge: Harvard University Press, 1987.

Lawrence, Christopher. "Incommunicable Knowledge: Science, Technology and the Clinical Art in Britain, 1850–1914." *Journal of Contemporary History* 20 (1985): 503–20

Lecuyer, Bernard-Pierre. "The Statistician's Role in Society: The Institutional Establishment of Statistics in France," *Minerva* 25 (1987): 35–55.

Lesch, John E. *Science and Medicine in France: The Emergence of Experimental Physiology, 1790–1855.* Cambridge: Harvard University Press, 1984.

Lilienfeld, A. M. "Ceteris paribus: The Evolution of the Clinical Trial." *Bulletin of the History of Medicine* 56 (1982): 1–18.

Llewelyn, D.E.H., and J. Anderson. "The Historical Development of the Concepts of Diagnosis and Prognosis and Their Relationship to Probabilistic Inference." *Medical Informatics* 5 (1980): 267–80.

MacKenzie, Donald A. *Statistics in Britain, 1865–1930: The Social Construction of Scientific Knowledge.* Edinburgh: Edinburgh University Press, 1981.

Marks, Harry Milton. "Ideas as Reforms: Therapeutic Experiments and Medical Practice, 1900–1980." Ph.D. dissertation, Massachusetts Institute of Technology, 1987.

Maulitz, Russell C. *Morbid Appearances: The Anatomy of Pathology in the Early Nineteenth Century.* Cambridge: Cambridge University Press, 1987.

Mayo, Deborah G., Rachelle D. Hollander, eds. *Acceptable Evidence: Science and Values in Risk Management.* Oxford: Oxford University Press, 1991.

Megill, Allan, ed. "Rethinking Objectivity, I & II." *Annals of Scholarship* 8 and 9 (1991–1992): 301–477; 1–153.

Menard, Claude. "Three Forms of Resistance to Statistics: Say, Cournot, Walras." *History of Political Economy* 12 (1980): 524–41.

Metivier, Michel, Pierre Costabel, and Pierre Dugac, eds. *Siméon-Denis Poisson et la science de son temps.* Palaiseau: Ecole Polytechnique, 1981.

Miké, Valerie. "Philosophers Assess Randomized Clinical Trials: The Need for Dialogue." *Controlled Clinical Trials* 10 (1989): 244–53.

Munk's Role. Lives of the Fellows of the Royal College of Physicians of London. Ed. R. R. Trail. Oxford: IRL Press at Oxford University Press, 1968, vol. 5. S.v. "Wright, Sir Almroth Edward."

Murphy, Terence D. "The French Medical Profession's Perception of Its Social Function between 1776 and 1830," *Medical History* 23 (1979): 259–78.

———. "Medical Knowledge and Statistical Methods in Early Nineteenth-Century France." *Medical History* 25 (1981): 301–19.

Osler, William. "Influence of Louis on American Medicine." *Bulletin of the Johns Hopkins Hospital* 8 (1897): 161–67.

Peset, José Luis, and Diego Gracia, eds. *The Ethics of Diagnosis.* Dordrecht: Kluwer, 1992.

Petit, L.-H. "Bulletin: Mort de M. le Professeur Gavarret." *Union médicale: Journal des intérêts scientifiques et practiques, moraux et professionnels du corps médical,* 3rd series, 50 (1890): 325.

Pickering, Andrew, ed. *Science as Practice and Culture.* Chicago: University of Chicago Press, 1992.

Pocock, Stuart J. *Clinical Trials: A Practical Approach.* New York: Wiley, 1983.

Polanyi, Michael. *Personal Knowledge.* Chicago: University of Chicago Press, 1957.

Porter, Theodore M. *The Rise of Statistical Thinking, 1820–1900.* Princeton: Princeton University Press, 1986.

———. "Objectivity and Authority: How French Engineers Reduced Public Utility to Numbers." *Poetics Today* 12 (Summer, 1991): 245–65.

Raigne-Delorme, Jacques. "Necrologie-M. Valleix." *Archives de médecine,* 5th series, 6 (1855): 243–50.

Reiser, Stanley Joel. *Medicine and the Reign of Technology.* Cambridge: Cambridge University Press, 1978.

Rolleston, Sir Humphry Davy. *The Right Honourable Sir Thomas Clifford Allbutt.* London: Macmillan, 1929.

Rosenberg, Charles E. *The Cholera Years.* Chicago: University of Chicago Press, 1987 [1962].

———. "Wood and Trees? Ideas and Actors in the History of Science." *Isis* 79 (1988): 565–70.

Rosenberg, Charles E., and Janet Golden, eds. *Framing Disease: Studies in Cultural History*. New Brunswick: Rutgers University Press, 1992.

"RU-486 Is on Its Way to the U.S." *The Washington Post*, 17 May, 1994, pp. A1, A4.

Rusnock, Andrea A. "On the Quantification of Things Human: Medicine and Political Arithmetic in Enlightenment England and France." Ph.D. dissertation, Princeton University, 1990.

Schiller, F. "The Statistician-Patient Relationship in Two Centuries: 'E Pluribus Unum.'" *International Congress of the History of Medicine* (24, Budapest, 1974). Acta 1976. Vol. 1, pp. 291–99.

Schiller, Joseph. "Claude Bernard et la statistique." *Archives internationales d'histoire des sciences* 16 (1963): 405–18.

Shapiro, Barbara J. *Probability and Certainty in Seventeenth-Century England*. Princeton: Princeton University Press, 1983.

Sheynin, Oskar B. "On the History of Medical Statistics." *Archive for History of Exact Sciences* 26 (1982): 241–80.

Shortt, S.E.D. "Physicians, Science, and Status: Issues in the Professionalization of Anglo-American Medicine in the Nineteenth Century." *Medical History* 27 (1983): 51–68.

Statistics in Medicine 1 (1982): 297–375 [devoted to career of A. B. Hill].

Staum, Martin S. *Cabanis: Enlightenment and Medical Philosophy in the French Revolution*. Princeton: Princeton University Press, 1980.

Stigler, Stephen M. *The History of Statistics: The Measurement of Uncertainty before 1900*. Cambridge: The Belknap Press of Harvard University Press, 1986.

Talbott, John H. *A Biographical History of Medicine*. New York: Grune and Stratton, 1970.

Tröhler, Ulrich. "Quantification in British Medicine and Surgery 1750–1830, with Special Reference to Its Introduction into Therapeutics." Ph.D. dissertation, University of London, 1978.

Turner, Frank M. *Between Science and Religion: The Reaction to Scientific Natualism in Late Victorian England*. New Haven: Yale University Press, 1974.

Turner, R. Steven. "The Growth of Professorial Research in Prussia, 1818–1848: Causes and Context." *Historical Studies in the Physical Sciences* 3 (1971): 137–82.

Turner, Thomas B. *Heritage of Excellence: The Johns Hopkins Medical Institutions, 1914–1947*. Baltimore: The Johns Hopkins University Press, 1974.

Vapereau, Gustave. *Dictionnaire universel des contemporains*, 4th ed. Paris: Hachette, 1870. S.v. "Dubois [d'Amiens], Frederick."

Vogel, Morris J., and Charles E. Rosenberg, eds. *The Therapeutic Revolution: Essays in the Social History of American Medicine*. Philadelphia: University of Pennsylvania Press, 1979.

Voit, C. "Adolph Fick (obituary)." *Sitzungsberichte der mathematisch-physikalischen der k. b. Akademie der Wissenschaften zu München* (1902): 277–87.

Von Furstenberg, George M., ed. *Acting Under Uncertainty: Multidiscipli-nary Conceptions.* Dordrecht: Kluwer, 1990.

Warner, John Harley. "Therapeutic Explanation and the Edinburgh Blood-letting Controversy: Two Perspectives on the Medical Meaning of Science in the Mid-Nineteenth Century." *Medical History* 24 (1980): 241–58.

———. *The Therapeutic Perspective: Medical Practice, Knowledge, and Iden-tity in America.* Cambridge: Harvard University Press, 1986.

———. "History of Science and History of Medicine." *Conference on Crit-ical Problems and Research Frontiers in History of Science and History of Technology.* Madison, Wisconsin: n.p., October 30–November 3, 1991.

———. "Ideals of Science and Their Discontents in Late-Nineteenth-Cen-tury American Medicine." *Isis* 82 (1991): 454–78.

INDEX

J. Rosser Matthews received his Ph.D. in the history of science from Duke University. In addition to teaching there, he has also taught at North Carolina State University and the University of Oklahoma.